Student Solutions Manual

Fundamentals of Algebraic Modeling
An Introduction to Mathematical Modeling with Algebra and Statistics

FIFTH EDITION

Daniel L. Timmons
Alamance Community College

Catherine W. Johnson
Alamance Community College

Sonya M. McCook
Alamance Community College

BROOKS/COLE
CENGAGE Learning

Australia • Brazil • Japan • Korea • Mexico • Singapore • Spain • United Kingdom • United States

© 2010 Brooks/Cole, Cengage Learning

ALL RIGHTS RESERVED. No part of this work covered by the copyright herein may be reproduced, transmitted, stored, or used in any form or by any means graphic, electronic, or mechanical, including but not limited to photocopying, recording, scanning, digitizing, taping, Web distribution, information networks, or information storage and retrieval systems, except as permitted under Section 107 or 108 of the 1976 United States Copyright Act, without the prior written permission of the publisher.

For product information and technology assistance, contact us at
**Cengage Learning Customer & Sales Support,
1-800-354-9706**

For permission to use material from this text or product, submit all requests online at **www.cengage.com/permissions**
Further permissions questions can be emailed to **permissionrequest@cengage.com**

ISBN-13: 978-0-495-55518-6
ISBN-10: 0-495-55518-5

Brooks/Cole
10 Davis Drive
Belmont, CA 94002-3098
USA

Cengage Learning is a leading provider of customized learning solutions with office locations around the globe, including Singapore, the United Kingdom, Australia, Mexico, Brazil, and Japan. Locate your local office at: **www.cengage.com/international**

Cengage Learning products are represented in Canada by Nelson Education, Ltd.

To learn more about Brooks/Cole, visit
www.cengage.com/brookscole

Purchase any of our products at your local college store or at our preferred online store
www.ichapters.com

Printed in the United States of America
1 2 3 4 5   14 13 12 11 10

Table of Contents

Chapter 1	**1**
Practice Set 1-2	1
Practice Set 1-3	2
Practice Set 1-4	6
Practice Set 1-5	9
Practice Set 1-6	13
Practice Set 1-7	15
Review Problems	18
Chapter 1 Test	24
Chapter 2	**29**
Practice Set 2-1	29
Practice Set 2-2	31
Practice Set 2-3	35
Practice Set 2-4	39
Practice Set 2-5	43
Review Problems	46
Chapter 2 Test	51
Chapter 3	**55**
Practice Set 3-1	55
Practice Set 3-2	57
Practice Set 3-3	59
Practice Set 3-4	62
Practice Set 3-5	66
Practice Set 3-6	70
Review Problems	71
Chapter 3 Test	78
Chapter 4	**83**
Practice Set 4-1	83
Practice Set 4-2	84
Practice Set 4-3	85
Practice Set 4-4	89
Practice Set 4-5	90
Practice Set 4-6	92
Practice Set 4-7	92
Practice Set 4-8	93
Review Problems	95
Chapter 4 Test	97
Chapter 5	**99**
Practice Set 5-1	99
Practice Set 5-2	102
Practice Set 5-3	105
Practice Set 5-4	106
Practice Set 5-5	107
Review Problems	108
Chapter 5 Test	110

Chapter 6   **113**
Practice Set 6-1   113
Practice Set 6-2   113
Practice Set 6-3   120
Practice Set 6-4   125
Review Problems   129
Chapter 6 Test   135

Chapter 7   **141**
Practice Set 7-1   141
Practice Set 7-2   142
Practice Set 7-3   143
Practice Set 7-4   146
Practice Set 7-5   147
Practice Set 7-6   149
Review Problems   152
Chapter 7 Test   154

Chapter 8   **157**
Practice Set 8-1   157
Practice Set 8-2   157
Practice Set 8-3   159
Practice Set 8-4   160
Practice Set 8-5   161
Practice Set 8-6   165
Review Problems   166
Chapter 8 Test   168

Chapter One

Practice Set 1-2

1. $\{-2.6, -\sqrt{5}, -1, 0, \frac{7}{13}, 3½, 4\}$

3. a) $\sqrt{7} < 3$ b) $-5 > -5¼$ c) $\pi > 0.3$

5. $5 + (-3) = 2$

7. $-6 + (-5) = -11$

9. $60 \div 0 =$ undefined (division by 0 is impossible)

11. $|6 - 8| = |-2| = 2$

13. $-1\frac{1}{3} + \frac{7}{8} = -\frac{4}{3} + \frac{7}{8} = -\frac{32}{24} + \frac{21}{24} = -\frac{11}{24}$

15. $(-2.5)(-1.6) = 4$

17. $\left(\frac{1}{4}\right)\left(-\frac{3}{8}\right) = -\frac{3}{32}$

19. $-(-2) + 10 = 2 + 10 = 12$

21. $-\frac{8}{9} \div \frac{4}{21} = -\frac{8}{9} \times \frac{21}{4} = -\frac{168}{36} = -\frac{168 \div 12}{36 \div 12} = -\frac{14}{3}$

23. $7 \cdot 3 - 2 \cdot 13 = 21 - 26 = -5$

25. $\dfrac{15}{[-16-(-11)]} = \dfrac{15}{-5} = -3$

27. $8 - (-6)(4 - 7) = 8 - (-6)(-3) = 8 - 18 = -10$

29. $\dfrac{2(-6+6)}{23-97} = \dfrac{2(0)}{-74} = \dfrac{0}{-74} = 0$

31. $(-8)^2 - 7(8) + 5 = 64 - 7(8) + 5 = 64 - 56 + 5 = 13$

33. $\dfrac{2^2 + 4^2}{5^2 - 3^2} = \dfrac{4+16}{25-9} = \dfrac{20}{16} = \dfrac{5}{4}$

35. $\dfrac{3(-5+1)}{12(3)+(-5+2)(-3-1)} = \dfrac{3(-4)}{12(3)+(-3)(-4)} = \dfrac{-12}{36+(12)} = \dfrac{-12}{48} = -\dfrac{1}{4}$

37. $\dfrac{5}{8} - 5\left(\dfrac{1}{8}\right) = \dfrac{5}{8} - \dfrac{5}{1}\left(\dfrac{1}{8}\right) = \dfrac{5}{8} - \dfrac{5}{8} = 0$

39. $1.25 + \dfrac{6.5}{0.5} + (0.25)^2 = 1.25 + \dfrac{6.5}{0.5} + 0.0625 = 1.25 + 13 + 0.0625 = 14.3125$

41. $\dfrac{3^3 - 2^3}{-4(-3+1)} = \dfrac{27-8}{-4(-2)} = \dfrac{19}{8} = 2.375$

43. -4[(-2)(6) - 7] = -4[(-12) - 7] = -4[-19] = 76

45. (3 - 8)(-2) - 10 = (-5)(-2) - 10 = 10 - 10 = 0

47. 230,000,000   (move decimal 8 places right)
49. 3000   (move decimal 3 places right)
51. 0.605   (move decimal 1 place left)
53. 0.003   (move decimal 3 places left)
55. 7.5   (move decimal 0 places)
57. 6.5×10^2
59. 3.2×10^9
61. 5×10^{-2}
63. 7.5×10^{-7}
65. 6.5×10^0
67. 1.47×10^8 km
69. 4.8×10^{-4}
71. 1.2495×10^{-9}
73. 4.375×10^{-10}
75. 4.6865×10^{21}

Practice Set 1-3

1.   x + 6 = -4
  x + 6 - 6 = -4 - 6
        x = -10

3.   3x + 7 = x
  3x + 7 - 7 = x - 7
    3x - x = x - 7 - x
      2x = -7
       x = $-\dfrac{7}{2}$ = -3.5

5.   2(x - 8) = 4
    2x - 16 = 4
  2x - 16 + 16 = 4 + 16
      2x = 20
       x = 10

7.   35 - 3x = 5
  35 - 3x - 35 = 5 - 35
      -3x = -30
       x = 10

9. $-4x + 5 = -54 + 41$
    $-4x + 5 = -13$
    $-4x + 5 - 5 = -13 - 5$
    $-4x = -18$
    $x = \dfrac{-18}{-4} = 4.5$

11. $7 - 3x = 2 - 5x$
    $7 - 3x - 7 = 2 - 5x - 7$
    $-3x = -5 - 5x$
    $-3x + 5x = -5 - 5x + 5x$
    $2x = -5$
    $x = \dfrac{-5}{2} = -2.5$

13. $3x + 9 = 42$
    $3x + 9 - 9 = 42 - 9$
    $3x = 33$
    $x = 11$

15. $-7(2x + 3) = -7$
    $-14x - 21 = -7$
    $-14x - 21 + 21 = -7 + 21$
    $-14x = 14$
    $x = -1$

17. $3(x - 2) + 2 = 11$
    $3x - 6 + 2 = 11$
    $3x - 4 = 11$
    $3x - 4 + 4 = 11 + 4$
    $3x = 15$
    $x = 5$

19. $3x + 5 = -5x - 8 + 1$
    $3x + 5 = -5x - 7$
    $3x + 5 + 5x = -5x - 7 + 5x$
    $8x + 5 = -7$
    $8x + 5 - 5 = -7 - 5$
    $8x = -12$
    $x = -\dfrac{12}{8} = -1.5$

21. $-\dfrac{3x}{8} = -\dfrac{15}{32}$
    $32\left(-\dfrac{3x}{8}\right) = 32\left(-\dfrac{15}{32}\right)$ [multiply by the common denominator 32]

$$-12x = -15$$
$$x = \frac{-15}{-12} = \frac{5}{4} = 1.25$$

23. $\quad \frac{x}{5} - 12 = 7$

   $5\left(\frac{x}{5}\right) - 5(12) = 5(7)$ [multiply by the common denominator 5]

   $x - 60 = 35$
   $x - 60 + 60 = 35 + 60$
   $x = 95$

25. $\quad w + ¾ = ⅝$
   $w + ¾ - ¾ = ⅝ - ¾$
   $w = ⅝ - 6/8 = -⅛$ [get a common denominator for ⅝ and ¾ and combine]

27. $\frac{2}{3}x + 8 = \frac{1}{2}$

   $6\left(\frac{2}{3}\right)x + 6(8) = 6\left(\frac{1}{2}\right)$ [multiply by the common denominator 6]

   $4x + 48 = 3$
   $4x + 48 - 48 = 3 - 48$
   $4x = -45$
   $x = \frac{-45}{4} = -11.25$

29. $\frac{1}{3} + \frac{1}{6}x - 2 = \frac{2}{3}x + \frac{1}{2}$

   $6\left(\frac{1}{3}\right) + 6\left(\frac{1}{6}x\right) - 6(2) = 6\left(\frac{2}{3}x\right) + 6\left(\frac{1}{2}\right)$ [multiply by the common denominator 6]

   $2 + x - 12 = 4x + 3$
   $x - 10 = 4x + 3$
   $x - 4x - 10 = 4x - 4x + 3$
   $-3x - 10 = 3$
   $-3x - 10 + 10 = 3 + 10$
   $-3x = 13$
   $x = -\frac{13}{3}$

31. $\quad \frac{3}{4}(4 - x) = 3 - x$

Fundamentals of Algebraic Modeling, 5th Ed.

$$3 - \frac{3}{4}x = 3 - x \qquad \text{[distributive property]}$$

$$4(3) - 4\left(\frac{3}{4}x\right) = 4(3) - 4(x) \qquad \text{[multiply through by the common denominator 4]}$$

$$12 - 3x = 12 - 4x$$
$$12 - 3x + 4x = 12 - 4x + 4x$$
$$x + 12 - 12 = 12 - 12$$
$$x = 0$$

33. $\quad$ k - 76.98 = 43.56
$\quad$ k - 76.98 + 76.98 = 43.56 + 76.98
$\qquad$ k = 120.54

35. x - 0.5x = 12
$\quad$ 0.5x = 12
$\quad$ x = 24

37. $\quad$ 3.5x + 3 = x - 1.75
$\quad$ 3.5x - x + 3 = x - x - 1.75
$\quad$ 2.5x + 3 = -1.75
$\quad$ 2.5x + 3 - 3 = -1.75 - 3
$\quad$ 2.5x = -4.75
$\quad$ x = -1.9

39. $\quad$ -4y + 3 = 12 - 4y $\qquad$ Since all of the variables have been eliminated
$\quad$ -4y + 4y + 3 = 12 - 4y + 4y $\qquad$ and the resulting statement (3 = 12)
$\qquad$ 3 = 12 $\qquad$ is false, there is no solution to the equation.
$\qquad$ ∅

41. $\quad$ 7a - (a - 5) = -10
$\quad$ 7a - a + 5 = -10
$\quad$ 6a + 5 = -10
$\quad$ 6a + 5 - 5 = -10 - 5
$\quad$ 6a = -15
$\quad$ $a = \dfrac{-15}{6} = -2.5$

43. 3(x - 5) - 5x = 2x + 9
$\quad$ 3x - 15 - 5x = 2x + 9
$\quad$ -2x - 15 = 2x + 9
$\quad$ -2x - 15 - 2x = 2x + 9 - 2x
$\quad$ -4x - 15 = 9
$\quad$ -4x - 15 + 15 = 9 + 15
$\quad$ -4x = 24
$\quad$ x = -6

45. $6x - 2(3 - 4x) = 6(3x + 2)$
    $6x - 6 + 8x = 18x + 12$
    $14x - 6 = 18x + 12$
    $14x - 18x - 6 = 18x - 18x + 12$
    $-4x - 6 = 12$
    $-4x - 6 + 6 = 12 + 6$
    $-4x = 18$
    $x = -4.5$

47. $4(3 - x) + 2x = -12$
    $12 - 4x + 2x = -12$
    $12 - 2x = -12$
    $12 - 2x - 12 = -12 - 12$
    $-2x = -24$
    $x = 12$

49. $3(x - 2) + 7 = 5 - 2(x + 3)$
    $3x - 6 + 7 = 5 - 2x - 6$
    $3x + 1 = -2x - 1$
    $3x + 1 - 1 = -2x - 1 - 1$
    $3x = -2x - 2$
    $3x + 2x = -2x + 2x - 2$
    $5x = -2$
    $x = -0.4$

Practice Set 1-4

1. $d = rt$
   $125 = 50t$ [Divide both sides by 50.]
   $t = 2.5$ hours

3. $I = Prt$
   $I = (\$5000)(0.05)(2) = \$500$

5. $A = \pi r^2$
   $A = \pi(3)^2 = (3.14)(9) = 28.26$ in^2

7. $A = \frac{1}{2}bh$
   $36 = \frac{1}{2}(8)h$
   $36 = 4h$
   9 in $= h$

9. $A = \sqrt{s(s-a)(s-b)(s-c)} = \sqrt{15(15-5)(15-12)(15-13)} =$
$\sqrt{15(10)(3)(2)} = \sqrt{900} = 30 \text{ in}^2$

11. $C = \dfrac{5}{9}(F - 32) = \dfrac{5}{9}(68 - 32) = \dfrac{5}{9}(36) = 20^\circ\, C$

13. $F = \dfrac{9}{5}C + 32 = \dfrac{9}{5}(-10) + 32 = -18 + 32 = 14^\circ\, F$

15. $m = \dfrac{y_2 - y_1}{x_2 - x_1} = \dfrac{-3-(-4)}{-1-2} = \dfrac{-3+4}{-1-2} = \dfrac{1}{-3} = -\dfrac{1}{3}$

17. $z = \dfrac{x - \bar{x}}{s} = \dfrac{95 - 100}{15} = \dfrac{-5}{15} = -\dfrac{1}{3}$

19. $a^2 + b^2 = c^2$
$3^2 + 4^2 = c^2$
$9 + 16 = c^2$
$25 = c^2$
$\sqrt{25} = 5 = c$

21. $M = P\left(1 + \dfrac{r}{n}\right)^{nt} = 5000\left(1 + \dfrac{0.045}{12}\right)^{12*10} = 5000\left(1 + \dfrac{0.045}{12}\right)^{120} = \$7834.96$

23. $x = \dfrac{-b \pm \sqrt{b^2 - 4ac}}{2a} = \dfrac{-5 \pm \sqrt{(5)^2 - 4(1)(-6)}}{2(1)} = \dfrac{-5 \pm \sqrt{25 + 24}}{2} =$

$\dfrac{-5 + \sqrt{49}}{2} = \dfrac{-5 + 7}{2} = \dfrac{2}{2} = 1$ or $\dfrac{-5 - \sqrt{49}}{2} = \dfrac{-5 - 7}{2} = \dfrac{-12}{2} = -6$

25. $y = Ae^{rn} = 1{,}500{,}000 e^{0.055*7} = 1{,}500{,}000 e^{0.385} = 2{,}204{,}421$ bacteria

27. $I = Prt$

$\dfrac{I}{Pt} = \dfrac{Prt}{Pt}$ [Divide both sides by Pt.]

$\dfrac{I}{Pt} = r$

29. $A = \dfrac{1}{2}bh$

$2A = bh$ [Multiply both sides by 2.]

$\dfrac{2A}{h} = b$ [Divide both sides by h to solve for b.]

31. $P = 2L + 2W$

$P - 2W = 2L + 2W - 2W$ [Subtract $2W$ from both sides.]

$P - 2W = 2L$

$\dfrac{P - 2W}{2} = L$ [Divide both sides by 2.]

33. $A = \tfrac{1}{2}(B + b)h$

$2(A) = 2[\tfrac{1}{2}(B + b)h]$ [Multiply both sides by 2.]

$2A = (B + b)h$

$2A = Bh + bh$ [Use the Distributive Property.]

$2A - bh = Bh + bh - bh$ [Subtract bh from both sides.]

$2A - bh = Bh$

$\dfrac{2A - bh}{h} = B$ [Divide both sides by h to solve for B.]

35. $2x + 3y = 6$

$2x + 3y - 2x = 6 - 2x$ [Subtract $2x$ from both sides.]

$3y = 6 - 2x$

$y = 2 - \tfrac{2}{3}x$ [Divide both sides by 3 to solve for y.]

Fundamentals of Algebraic Modeling, 5th Ed.

37. $$A = \frac{x+y}{2}$$
$2A = x + y$ [Multiply both sides by *2*.]
$2A - y = x + y - y$ [Subtract *y* from both sides.]
$2A - y = x$

39. $$F = \frac{9}{5}C + 32$$
$5F = 9C + 160$ [Multiply through by 5.]
$5F - 160 = 9C + 160 - 160$ [Subtract 160 from both sides.]
$5F - 160 = 9C$
$\frac{5}{9}F - \frac{160}{9} = C$ [Divide both sides by 9.]
$\frac{5}{9}(F - 32) = C$ [Factor out $\frac{5}{9}$.]

Practice Set 1-5

1. $\frac{45\ min}{2\ hr} = \frac{45\ min}{120\ min} = \frac{3}{8}$

3. $\frac{4\ in}{4\ ft} = \frac{4\ in}{48\ in} = \frac{1}{12}$

5. $\frac{6\ ft}{3\ yd} = \frac{6\ ft}{9\ ft} = \frac{2}{3}$

7. $\frac{8\ weeks}{16\ days} = \frac{56\ days}{16\ days} = \frac{7}{2}$

9. $\frac{25\ mL}{1\ L} = \frac{25\ mL}{1000\ mL} = \frac{1}{40}$

11. $\frac{304\ miles}{9.5\ gal} = 32$ mi./gal.

13. $\frac{\$3.50}{10\ min} = \$0.35$/min.

Fundamentals of Algebraic Modeling, 5th Ed.

15. $\dfrac{\$48}{10\,days} = \$4.80/\text{day}$

17. $\dfrac{24\,lb}{15\,people} = 1.6 \text{ lb./person}$

19. $\dfrac{x}{5} = \dfrac{3}{4}$
    $4x = (3)(5)$ [cross-multiplication property]
    $4x = 15$
    $x = \dfrac{15}{4}$

21. $\dfrac{30}{126} = \dfrac{5}{3x}$
    $(3x)(30) = 5(126)$ [cross-multiplication property]
    $90x = 630$
    $x = 7$

23. $\dfrac{3x+6}{35} = \dfrac{2x-18}{5}$
    $5(3x + 6) = 35(2x - 18)$ [cross-multiplication property]
    $15x + 30 = 70x - 630$ [distributive property]
    $15x + 30 - 70x = 70x - 630 - 70x$
    $-55x + 30 = -630$
    $-55x + 30 - 30 = -630 - 30$
    $-55x = -660$
    $x = 12$

25. $\dfrac{15}{18} = \dfrac{x-1}{x}$
    $(15)(x) = 18(x - 1)$ [cross-multiplication property]
    $15x = 18x - 18$ [distributive property]
    $15x - 18x = 18x - 18 - 18x$
    $-3x = -18$
    $x = 6$

27. $\dfrac{3}{x+1} = \dfrac{18}{9x-3}$
    $3(9x - 3) = 18(x + 1)$ [cross-multiplication property]
    $27x - 9 = 18x + 18$ [distributive property]
    $27x - 9 - 18x = 18x + 18 - 18x$
    $9x - 9 = 18$
    $9x - 9 + 9 = 18 + 9$
    $9x = 27$
    $x = 3$

Fundamentals of Algebraic Modeling, 5th Ed.

29. Unit rate equals cost divided by the number of square feet.
    $$\$2235 \div 1500 \text{ ft}^2 = \$1.49/\text{ft}^2$$

31. Cost per ounce = $\$8.99 \div 16$ oz = $\$0.561875$ per ounce

33. Brand X: $\$1.49 \div 8$ oz = $\$0.18625$/ounce
    Brand Z: $\$2.12 \div 12$ oz = $\$0.1766\ldots$/ounce
    Brand Z is the better buy because it costs less per ounce.

35. Let $\dfrac{120 \text{ cal}}{\sfrac{3}{4} \text{ cup}} = \dfrac{x}{1 \text{ cup}}$.

    $(1)(120) = \tfrac{3}{4}x$ [Cross multiply.]
    $120 = \tfrac{3}{4}x$
    $160 \text{ calories} = x$ [Multiply both sides by 4/3.]

37. Let $\dfrac{2.5 \text{ dozen}}{1.25 \text{ cups}} = \dfrac{x \text{ dozen}}{3 \text{ cups}}$.

    $(3)(2.5) = 1.25x$ [Cross multiply.]
    $7.5 = 1.25x$
    $6 = x$
    Answer is 6 dozen or 72 muffins.

39. Let $\dfrac{5 \text{ ft } 4 \text{ in}}{10.5 \text{ ft}} = \dfrac{x}{20 \text{ ft}}$. Convert the measurements to inches as follows: 5 ft 4 in = 5(12) + 4 = 64 inches; 10.5 ft = 10.5 × 12 in = 126 inches; 20 ft = 20 × 12 in = 240 inches. Now substitute:

    $$\dfrac{64 \text{ in}}{126 \text{ in}} = \dfrac{x}{240 \text{ in}}$$
    $(64)(240) = 126x$
    $15360 = 126x$
    $121.9 \text{ inches} = x$

    Converting to feet and inches: 121.9 inches ÷ 12in/ft = 10 ft. 1.9 in.

41. Let $\dfrac{1.8 A}{18 V} = \dfrac{5.4 A}{x}$.

    $(18)(5.4) = 1.8A$ [Cross multiply.]
    $97.2 = 1.8A$
    $54 V = A$

43. Let $\dfrac{110 \text{ lb}}{19.4 \text{ lb}} = \dfrac{200 \text{ lb}}{x}$

    $110x = (19.4)(200)$ [Cross multiply.]
    $110x = 3880$
    $x = 35.2727\ldots$ or approximately 35.3 lb.

45. Let $\dfrac{10\,lb}{400\,ft^2} = \dfrac{x}{500\,ft^2}$

$(10)(500) = 400x$ [Cross multiply.]
$5000 = 400x$
$12.5\ lb = x$

47. Let $\dfrac{\$240}{1200\,words} = \dfrac{x}{1500\,words}$.

$(240)(1500) = 1200x$ [Cross multiply.]
$360{,}000 = 1200x$
$\$300 = x$

49. Let $\dfrac{1\,adult}{15\,children} = \dfrac{3\,adults}{x}$

$x = (3)(15)$ [Cross multiply.]
$x = 45\ children$

51. Let $\dfrac{1\,inch}{8\,feet} = \dfrac{2.75\,inches}{x}$

$x = (8)(2.75)$ [Cross multiply.]
$x = 22\ feet$

Let $\dfrac{1\,inch}{8\,feet} = \dfrac{1.9375\,inches}{x}$

$x = (8)(1.9375)$ [Cross multiply.]
$x = 15.5\ feet$
Dimensions are 22 ft × 15.5 ft

53. Let $\dfrac{Mach\,1}{761.2\,mph} = \dfrac{Mach\,3.1}{x}$

$1x = (3.1)(761.2)$ [Cross multiply.]
$x = 2359.72\ mph$

55. Let $\dfrac{400\,ft}{1\,day} = \dfrac{1000\,ft}{x}$

$(1)(1000) = 400x$ [Cross multiply.]
$1000 = 400x$
$2.5\ days = x$

57. Let $\dfrac{40\,mg}{1\,mL} = \dfrac{60\,mg}{x}$.

$40x = (1)(60)$ [Cross multiply.]
$40x = 60$
$x = 1.5\ mL$

Fundamentals of Algebraic Modeling, 5th Ed.

59. Let $\dfrac{1.8\,mi}{30\,\min} = \dfrac{x}{45\,\min}$

$(1.8)(45) = 30x$ [Cross multiply.]

$81 = 30x$

$2.7 \text{ miles} = x$

Practice Set 1-6

1. $8\% = \dfrac{8}{100} = \dfrac{2}{25}$

3. $180\% = \dfrac{180}{100} = \dfrac{9}{5}$

5. $33\dfrac{1}{3}\% = \dfrac{100}{3}\% = \dfrac{\frac{100}{3}}{100} = \dfrac{100}{3} \div 100 = \dfrac{100}{3} \cdot \dfrac{1}{100} = \dfrac{1}{3}$

7. $5\% = 5 \div 100 = 0.05$

9. $0.05\% = 0.05 \div 100 = 0.0005$

11. $1.5\% = 1.5 \div 100 = 0.015$

13. $125\% = 125 \div 100 = 1.25$

15. $0.005 \cdot 100 = 0.5$ so $0.005 = 0.5\%$

17. $\dfrac{2}{3} \cdot 100 = \dfrac{200}{3} = 66\dfrac{2}{3}$ so $\dfrac{2}{3} = 66\dfrac{2}{3}\%$

19. $1.5 \cdot 100 = 150$ so $1.5 = 150\%$

21. 15% of $75 = 0.15 \times 75 = 11.25$

23. 1.5% of $32 = 0.015 \times 32 = 0.48$

25. 2.5% of $x = 33.5$

$0.025x = 33.5$

$x = 1340$

27. 15% of $x = 7.5$

$0.15x = 7.5$

$x = 50$

29. 306 is what percent of 450?

$306 = x(450)$

$\dfrac{306}{450} = x$

$0.68 = x$

$68\% = x$

Fundamentals of Algebraic Modeling, 5th Ed.

Using proportions, $\frac{x}{100} = \frac{306}{450}$

$450x = (306)(100)$

$x = 68$ or 68%

31. 0.375 is what percent of 25?

    $0.375 = x(25)$

    $\frac{0.375}{25} = x$

    $0.015 = x$

    $1.5\% = x$

    Using proportions, $\frac{x}{100} = \frac{0.375}{25}$

    $25x = (0.375)(100)$

    $25x = 37.5$

    $x = 1.5$ or 1.5%

33. $\frac{x}{100} = \frac{18 \text{ free throws}}{24 \text{ attempts}}$

    $24x = (100)(18)$

    $24x = 1800$

    $x = 75$ or 75%

35. 65% of free throws = $0.65 \times 12 = 7.8$ or 8 free throws.

37. 18% of the bill = $0.18 \times \$325 = \$58.50$

39. 26% of weight = $0.26 \times 180 = 46.8$ lb. of skin

41. 9% of purchase price = $0.09 \times \$42.50 = \$3.83$

43. $\frac{x}{100} = \frac{6}{200}$

    $200x = (6)(100)$

    $200x = 600$

    $x = 3$ or 3% sleep more than 8 hours

    3% of people = $0.03 \times 700 = 21$ people out of 700 sleep more than 8 hours

45. 8.3% of Georgia residents aged 25 and older = $0.083 \times 5{,}185{,}965 = 430{,}435$ have professional or graduate degrees.

47. 4% increase × current pay = $0.04 \times \$7.75 = \$0.31$ increase;
    new rate = $\$7.75 + \$0.31 = \$8.06$

49. $\dfrac{151.80 - 172.50}{172.50} \cdot 100 = \dfrac{-20.7}{172.50} \cdot 100 = -12$ or a discount of 12%

51. $\dfrac{\$80,000 - \$75,000}{\$75,000} \cdot 100 = \dfrac{\$5,000}{\$75,000} \cdot 100 = 6.\overline{6}$ or approximately $6\dfrac{2}{3}\%$ increase

Practice Set 1-7

1. 2x + 6

3. 7x – 2

5. 3(4 + x)

7. $\dfrac{1}{3}x - 5$

9. Let Bill's salary = x.   Then, Ann's salary = x + $5000

11. Let the width = x. Then, the length = 2x + 5.

13. 5x + 5 = 2x – 10
 5x + 5 – 2x = 2x – 10 – 2x
 3x + 5 = -10
 3x + 5 – 5 = -10 – 5
 3x = -15
 x = -5

15. Emily has saved x. Elena has saved 2x (twice as much).
 x + 2x = 72
 3x = 72
 x = 24 so Emily has saved $24 and Elena has saved 2($24) = $48.

17. Let x = the shorter piece. The longer piece will be 2x.
 x + 2x = 60 meters
 3x = 60 meters
 x = 20 meters
 The shorter piece is 20 meters long and the longer piece is 2(20) = 40 meters long.

19. Using the definition of an average, let x equal the missing grade.
 $\dfrac{88 + 91 + 95 + x}{4} = 90$
 88 + 91 + 95 + x = 4(90)
 274 + x = 360
 274 - 274 + x = 360 - 274
 x = 86
 In order to have a 90 average, she must make 86 on the last test.

Fundamentals of Algebraic Modeling, 5th Ed.

21. Let x = the number of kilowatt-hours. Write the equation and solve.
$$\$20.00 + \$0.14x = \$85.78$$
$$20.00 - 20.00 + 0.14x = 85.78 - 20.00$$
$$0.14x = 65.78$$
$$x = \frac{65.78}{0.14} = 470 \; kWh \quad \text{(rounded)}$$

23. Let x = the taxi fare. Since 1/11 of a mile costs $0.20, the cost of one mile = $0.20 × 11 = $2.20.
$$x = \$3.20 + 18(\$2.20/mile) + \$1.20 = \$44.00$$

25. The opponent scored x points in the game. Mighty Mites scored 39 points. This amount (39) is 2x - 1.
$$2x - 1 = 39$$
$$2x - 1 + 1 = 39 + 1$$
$$2x = 40$$
$$x = 20$$
The opponent scored 20 points.

27. Let the integers be x, x + 1 and x + 2. Write the equation for the sum and solve.
$$x + (x + 1) + (x + 2) = 87$$
$$3x + 3 = 87$$
$$3x + 3 - 3 = 87 - 3$$
$$3x = 84$$
$$x = 28$$
The integers are x = 28, x + 1 = 29 and x + 2 = 30.

29. Let the odd integers be x, x + 2, and x + 4. Write the equation for the sum and solve.
$$x + (x + 2) + (x + 4) = -273$$
$$3x + 6 = -273$$
$$3x + 6 - 6 = -273 - 6$$
$$3x = -279$$
$$x = -93$$
The three integers are x = -93, x + 2 = -91, and x + 4 = -89.

31. Let x = the value of the lot. Then, 6.5x = the value of the house.
$$x + 6.5x = \$175,000$$
$$7.5x = \$175,000$$
$$x = \$23,333.33$$
So, the lot is worth approximately $23,333 and the house is worth about $151,667.

33. $x = \dfrac{949 \; yen}{146 \; yen/dollar} = \$6.50.$

Fundamentals of Algebraic Modeling, 5th Ed.

35. $x = 500$ miles $\div 151.744$ mph $= 3.295$ hours. Convert 0.295 to minutes by multiplying $0.295(60) = 17.7$ to give the answer of approximately 3 hours and 18 minutes.

37. Let the number of males and females at the beginning of the semester $= x$.
    If 8 males drop the class, there are $x - 8$ males remaining. Now the number of females is twice the number of males remaining.
    $$x = 2(x - 8)$$
    $$x = 2x - 16$$
    $$x - 2x = 2x - 16 - 2x$$
    $$-x = -16$$
    $$x = 16$$
    There were 16 males and 16 females at the beginning of the semester.

39. Let $x =$ the number of bags of apples.
    $$5x + 2x = 252$$
    $$7x = 252$$
    $$x = 36$$
    There are 36 bags containing 5 lb. of apples and 36 bags containing 2 lb. of apples.

41. Let $x =$ profit from the automotive division. Then $x - 273$ represents the profit from financial services.
    $$x + (x - 273 \text{ million}) = 483 \text{ million}$$
    $$2x - 273 \text{ million} = 483 \text{ million}$$
    $$2x - 273 \text{ million} + 273 \text{ million} = 483 \text{ million} + 273 \text{ million}$$
    $$2x = 756 \text{ million}$$
    $$x = 378 \text{ million}$$

    The profit from the automotive division was 378 million and from financial services was 378 million – 273 million = 105 million.

43. Let $x =$ attendance at the Ohio State game and $x + 2471 =$ attendance at the Penn State game. Total attendance for the two teams in 2006 was 212,663.
    $$x + x + 2471 = 212{,}663$$
    $$2x + 2471 = 212{,}663$$
    $$2x + 2471 - 2471 = 212{,}663 - 2471$$
    $$2x = 210{,}192$$
    $$x = 105{,}096$$
    Therefore, the attendance at the Ohio State game was 105,096 and the attendance at the Penn State game was 107,567.

45. Let $x =$ the original price of the radio. Savings $= 15\%$ of retail price or $0.15x$.
    $$x - 0.15x = \$127.46$$
    $$0.85x = \$127.46$$
    $$x = \$149.95$$
    The original price of the radio was $149.95.

47. Let x = the wholesale price of the shoes. The markup amount is 65% times x.
$$x + 0.65x = \$125.40$$
$$1.65x = \$125.40$$
$$x = \$76$$
The wholesale price of the shoes was $76.00.

49. Let x = Drema's contributions. The company's contributions = 20% of x.
$$x + 0.20x = \$1200$$
$$1.2x = \$1200$$
$$x = \$1000$$
Drema deposited $1000 into the account.

Chapter 1 Review Problems
1. $-\dfrac{2}{3}, \dfrac{3}{2}$
2. $9, -\dfrac{1}{9}$
3. $-|-3 + 5| = -|2| = -2$
4. $-7^2 + 9 = -49 + 9 = -40$
5. $6 - 3(4 - 7) \div 4 = 6 - 3(-3) \div 4 = 6 - (-9) \div 4 = 6 - (-2.25) = 8.25$
6. $23.89 \div (-3.48) = -6.8649425\ldots$ or -6.865 (rounded)
7. $3^2 + |-28| - (6^2 - 8) = 9 + 28 - (36 - 8) = 9 + 28 - 28 = 9$
8. $-\dfrac{2}{3} \div \dfrac{4}{5} - \left(-\dfrac{1}{2}\right) = -\dfrac{2}{3} \cdot \dfrac{5}{4} + \dfrac{1}{2} = -\dfrac{5}{6} + \dfrac{3}{6} = -\dfrac{2}{6} = -\dfrac{1}{3}$
9. 42,500
10. 510
11. 0.0000175
12. 0.61
13. 1.508×10^6
14. 8.5×10^0
15. 2.78×10^{-1}
16. 1.08×10^{-4}

17. $$2x - 9 = -5x + 7$$
$$2x - 9 + 5x = -5x + 7 + 5x$$
$$7x - 9 = 7$$
$$7x - 9 + 9 = 7 + 9$$
$$7x = 16$$
$$x = \dfrac{16}{7} = 2\dfrac{2}{7}$$

18. $$2x - 7x = 15$$
$$-5x = 15$$
$$x = -3$$

19. $3 = \dfrac{1}{3} - 2x$

$3(3) = 3\left(\dfrac{1}{3}\right) - 3(2x)$

$9 = 1 - 6x$

$9 - 1 = 1 - 6x - 1$

$8 = -6x$

$\dfrac{8}{-6} = -\dfrac{4}{3} = x$

20.   -5(n - 2) = 8 - 4n
   -5n + 10 = 8 - 4n
   -5n + 10 + 4n = 8 - 4n + 4n
   -n + 10 = 8
   -n + 10 - 10 = 8 - 10
   -n = -2
   n = 2

21. $\dfrac{2x}{3} - 5 = 7$

$3\left(\dfrac{2x}{3}\right) - 3(5) = 3(7)$

2x - 15 = 21
2x - 15 + 15 = 21 + 15
2x = 36
x = 18

22. 2x + 3(x − 5) = 15
   2x + 3x − 15 = 15
   5x − 15 = 15
   5x - 15 + 15 = 15 + 15
   5x = 30
   x = 6

23. -2(x - 1) - 3x = 8 - 4x
   -2x + 2 - 3x = 8 - 4x
   -5x + 2 + 4x = 8 - 4x + 4x
   -x + 2 - 2 = 8 - 2
   -x = 6
   x = -6

24. 7(x - 2) - 6(x + 1) = -20
   7x - 14 - 6x - 6 = -20
   x - 20 = -20
   x - 20 + 20 = -20 + 20
   x = 0

25. $I = Prt$
    $t = \dfrac{I}{Pr}$ [Divide both sides by Pr.]

26. $2x + 3y = 9$
    $3y = -2x + 9$ [Subtract $2x$ from both sides.]
    $y = -\dfrac{2}{3}x + 3$ [Divide all terms by 3.]

27. $C = \pi d$
    $d = \dfrac{C}{\pi}$ [Divide both sides by π.]

28. $a + b + c = P$
    $c = P - a - b$ [Subtract a and b from both sides.]

29. $\dfrac{27 \, min}{3 \, hr} = \dfrac{27 \, min}{180 \, min} = \dfrac{3}{20}$

30. $\dfrac{4 \, weeks}{21 \, days} = \dfrac{28 \, days}{21 \, days} = \dfrac{4}{3}$

31. $\dfrac{6 \, in}{5 \, ft} = \dfrac{6 \, in}{60 \, in} = \dfrac{1}{10}$

32. $\dfrac{\$60}{5 \, hr} = \$12/\text{hour}$

33. $\dfrac{44 \, bushels}{8 \, trees} = 5.5 \, \text{bushels/tree}$

34. $\dfrac{\$12.80}{3.5 \, lb} = \$3.66/\text{lb.}$

35. $\dfrac{x}{3} = \dfrac{4}{7}$
    $7x = 12$
    $x = \dfrac{12}{7}$

36. $\dfrac{2}{3} = \dfrac{8}{2x}$
    $4x = 24$
    $x = 6$

37. $\dfrac{x-3}{8} = \dfrac{3}{4}$
   4x - 12 = 24
   4x - 12 + 12 = 24 + 12
   4x = 36
   x = 9

38. $\dfrac{4x-3}{7} = \dfrac{2x-1}{3}$
   12x - 9 = 14x - 7
   12x - 9 - 14x = 14x - 7 - 14x
   -2x - 9 + 9 = -7 + 9
   -2x = 2
   x = -1

39. 12.5% of 150 = 0.125(150) = 18.75

40. x(80) = 10
   $x = \dfrac{10}{80} = 0.125$
   0.125 × 100 = 12.5 so x = 12.5%

41. 4%(x) = 64
   0.04x = 64
   $x = \dfrac{64}{0.04} = 1600$

42. $I = Prt$. Substitute: $I = (\$2000)(0.08)(2) = \$320$

43. $A = \tfrac{1}{2}bh$. Substitute: $A = \tfrac{1}{2}(3 \text{ in})(4 \text{ in}) = 6 \text{ in}^2$

44. Let $\dfrac{2400\,L}{50\,min} = \dfrac{x}{30\,min}$. Cross multiply.
   72000 = 50x
   1440 liters = x

45. Let $\dfrac{3}{4} = \dfrac{x}{92}$. Cross multiply.
   276 = 4x
   69 dentists = x

46. Let $\dfrac{76.8\,lb\ copper}{160\,lb\ alloy} = \dfrac{x}{100}$. Cross multiply.
   7680 = 160x
   x = 48 so there is 48% copper

Fundamentals of Algebraic Modeling, 5th Ed.

47. $(1.5\%)(\$1800) = 0.015(1800) = \$27$ raise; New salary $= \$1800 + \$27 = \$1827$

48. Let $\dfrac{80}{100} = \dfrac{x}{2.00\ kg}$. Cross multiply.
    $$160 = 100x$$
    $$1.6\ kg = x$$

49. Calculate the expansion by multiplying $0.45\%(1.5671)$ and adding to the original length. $(0.45\%)(1.5671) = (0.0045)(1.5671) = 0.00705195$. Add this to 1.5671 to get an expanded length of 1.57415195 or 1.5742 meters (rounded).

50. $\dfrac{50-46.1}{50} \cdot 100 = \dfrac{3.9}{50} \cdot 100 = 7.8$ so 7.8% decrease in volume

51. $\dfrac{0.81-0.75}{0.75} \cdot 100 = \dfrac{0.06}{0.75} \cdot 100 = 8$ so 8% price increase

52. $\dfrac{\$5-\$4}{\$4} \cdot 100 = \dfrac{1}{4} \cdot 100 = 25$, so 25% increase; Applying this to a $9 former charge for a suit gives $\$9 + 0.25(\$9) = \$9 + \$2.25 = \$11.25$.

53. Let x = total number of minutes that you talk. Then, $\$1.24 + (x - 4)(\$0.28) = \$3.76$
    $$1.24 + 0.28x - 1.12 = 3.76$$
    $$0.12 + 0.28x = 3.76$$
    $$0.12 + 0.28x - 0.12 = 3.76 - 0.12$$
    $$0.28x = 3.64$$
    $$x = 13 \quad \text{The call was 13 minutes long.}$$

54. Let x = number of minutes per month. Then, $\$50 + \$0.36x = \$99.68$.
    $$0.36x = 99.68 - 50$$
    $$0.36x = 49.68$$
    $$x = 138 \text{ minutes}$$

55. Let x = labor time (per hour). Then, $\$40 + \$30x = \$115$
    $$40 + 30x - 40 = 115 - 40$$
    $$30x = 75$$
    $$x = 2.5 \text{ hours}$$

56. Let the shorter piece be x and the longer be $x + 3$. Then, $x + (x + 3) = 33$
    $$2x + 3 = 33$$
    $$2x + 3 - 3 = 33 - 3$$
    $$2x = 30$$
    $$x = 15 \text{ in (short piece)} \quad \text{and} \quad x + 3 = 15 + 3 = 18 \text{ in (long piece)}$$

57. Let the short piece = x, the middle-sized piece = $x + 14$, and the long piece = $3x$.
Then, $x + (x + 14) + 3x = 79$
$5x + 14 = 79$
$5x + 14 - 14 = 79 - 14$
$5x = 65$
$x = 13$ cm (short)
$x + 14 = (13) + 14 = 27$ cm (middle)
$3x = 3(13) = 39$ cm (long)

58. Let the calculator be x and the cassette player be $x + 140$.
Then, $x + (x + 140) = 208$
$2x + 140 = 208$
$2x + 140 - 140 = 208 - 140$
$2x = 68$
$x = \$34.00$ which is the cost of the calculator

59. Let n = the number of nickels and $2n - 2$ = the number of dimes.
Then, $n + (2n - 2) = 52$.
$3n - 2 = 52$
$3n - 2 + 2 = 52 + 2$
$3n = 54$
$n = 18$ There are 18 nickels and 34 dimes in the bank for a total of $4.30.

60. Use the formula $d = rt$ and substitute values.
$182 \text{ mi} = (52 \text{ mph})(t)$
$t = 3.5$ hours

61. Let x = the original price of the ring. Savings = 40% of original price.
$x - 0.40x = \$599.99$
$0.60x = \$599.99$
$x = \$999.98$

62. Let x = her original hourly wage. Increase = 10% of original hourly wage.
$x + 0.10x = \$8.25$
$1.10x = \$8.25$
$x = \$7.50$

63. Let x = your friend's former weight. Loss = 12% of her former weight.
$x - 0.12x = 110$
$0.88x = 110$
$x = 125$ lb.

Fundamentals of Algebraic Modeling, 5th Ed.

64. $T = UN + F$ Substitute the given values and solve.
$$\$16,750 = \$15N + \$2500$$
$$16750 - 2500 = 15N + 2500 - 2500$$
$$14250 = 15N$$
$$N = 950 \text{ units produced}$$

65. First remove the parentheses by use of the distributive property, then add 2 to the 12.
$$2 + 3(2x + 4) = 2 + 6x + 12 = 6x + 14.$$

Chapter 1 Test

1. The absolute value of a number is its distance from zero on the number line, and distances are always positive. Both 8 and -8 are eight units from zero on the number line.

2. It is the "same" number with the opposite sign or, it is the number with the same absolute value but on the opposite side of zero on the number line.

3. $2(5 - 7)^3 = 2(-2)^3 = 2(-8) = -16$

4. $|3^2 - 2(6 - 5)| = |3^2 - 2(1)| = |9 - 2| = 7$

5. $\dfrac{6(-5)}{2} + 3(6 - 9) - 7 = \dfrac{-30}{2} + 3(-3) - 7 = -15 + (-9) - 7 = -31$

6. 0.000208

7. 6.12×10^9

8.
$$3(y + 7) = 2y - 5$$
$$3y + 21 = 2y - 5$$
$$3y + 21 - 2y = 2y - 5 - 2y$$
$$y + 21 = -5$$
$$y + 21 - 21 = -5 - 21$$
$$y = -26$$

9.
$$\dfrac{5x}{6} = 2x - 7$$
$$6\left(\dfrac{5x}{6}\right) = 6(2x) - 6(7)$$
$$5x = 12x - 42$$
$$5x - 12x = 12x - 42 - 12x$$
$$-7x = -42$$
$$x = 6$$

10. $6x + x - 0.9 = 6x + 0.9$
    $7x - 0.9 = 6x + 0.9$
    $7x - 0.9 - 6x = 6x + 0.9 - 6x$
    $x - 0.9 = 0.9$
    $x - 0.9 + 0.9 = 0.9 + 0.9$
    $x = 1.8$

11. $4y - 6(y + 4) = 1 - y$
    $4y - 6y - 24 = 1 - y$
    $-2y - 24 = 1 - y$
    $-2y - 24 + y = 1 - y + y$
    $-y - 24 = 1$
    $-y - 24 + 24 = 1 + 24$
    $-y = 25$
    $y = -25$

12. $7 - (3x - 1) = 5x$
    $7 - 3x + 1 = 5x$
    $8 - 3x = 5x$
    $8 - 3x + 3x = 5x + 3x$
    $8 = 8x$
    $1 = x$

13. $V = lwh$
    $w = \dfrac{V}{lh}$ [Divide both sides by lh.]

14. $h = vt - 16t^2$
    $h + 16t^2 = vt$ [Add $16t^2$ to both sides.]
    $\dfrac{h + 16t^2}{t} = v$ [Divide both sides by t.]

15. $\dfrac{4\ hr}{1\ day} = \dfrac{4\ hr}{24\ hr} = \dfrac{1}{6}$

16. $\dfrac{10\ ft}{160\ in} = \dfrac{120\ in}{160\ in} = \dfrac{3}{4}$

17. $\dfrac{\$413.20}{4\ days} = \$103.30 / day$

18. $\dfrac{7.5\ lb}{6\ people} = 1.25\ lb / person$

19. $\dfrac{50 \text{ eggs}}{10 \text{ chickens}} = 5 \text{ eggs / chicken}$

20. $\dfrac{7}{12} = \dfrac{3x}{10}$
    $70 = 36x$
    $\dfrac{70}{36} = \dfrac{35}{18} = x$

21. $\dfrac{x-4}{8} = \dfrac{2x+3}{9}$
    $9(x - 4) = 8(2x + 3)$
    $9x - 36 = 16x + 24$
    $9x - 36 - 16x = 16x + 24 - 16x$
    $-7x - 36 = 24$
    $-7x - 36 + 36 = 24 + 36$
    $-7x = 60$
    $x = -\dfrac{60}{7}$

22. 15% of 85 = 0.15 × 85 = 12.75

23. Using proportions: $\dfrac{x}{100} = \dfrac{33.5}{150}$
    $150x = 3350$
    $x = \dfrac{3350}{150} = 22.3\overline{3}$ *so approximately* 22.3%

24. Using proportions: $\dfrac{2.5}{100} = \dfrac{16.25}{x}$
    $2.5x = 1625$
    $x = 650$

25. Let x = sales tax.
    Then x = (6%)($29.95) = (0.06)(29.95) = $1.80

26. Let x = tip. If 15% of *the cost of the meal* is the *tip*, then
    (0.15)($45.80) = x
    $6.87 = x

27. Let x = defective bulbs. Then, $\dfrac{3}{85} = \dfrac{x}{510}$.
    $85x = 1530$
    $x = 18$ bulbs

28. Let x = parts produced. Then, $\dfrac{300 \ parts}{20 \ min} = \dfrac{x}{45 \ min}$.
$$20x = 13500$$
$$x = 675 \ parts$$

29. Given $P = 2L + 2W$. Substitute the given values into the formula.
$$P = 2(20 \ ft.) + 2(12 \ ft.) = 64 \ ft.$$

30. Company A Plan: $\$20 + 0.10m$ \qquad (where m = miles driven)
Company B Plan: $\$10 + 0.30m$
To find the number of miles where the two costs are equal, set these two expressions equal.
$$\$20 + 0.10m = \$10 + 0.30m$$
$$20 + 0.10m - 0.10m = 10 + 0.30m - 0.10m$$
$$20 = 10 + 0.20m$$
$$20 - 10 = 10 + 0.20m - 10$$
$$10 = 0.20m$$
$$50 = m, \text{ so at 50 miles the costs are the same for both plans}$$

31. % decrease $= \dfrac{\$6500 - \$15{,}000}{\$15{,}000} = \dfrac{-\$8500}{\$15{,}000} = -0.56666 = -56.7\%$ or 57% decrease

32. Let the number of passengers on one ship = x.
The second ship holds twice as many passengers = $2x$
Write the equation $x + 2x = 2250$ and solve.
$$3x = 2250$$
$$x = 750 \text{ so the smaller ship holds 750 passengers}$$

33. Let Sarah's age be x. Then, Michelle's age is $5x - 10$. Write the equation and solve.
$$x + 5x - 10 = 44$$
$$6x - 10 = 44$$
$$6x - 10 + 10 = 44 + 10$$
$$6x = 54$$
$$x = 9 \quad \text{so Sarah is 9 years and Michelle is } 5(9)-10 = 35 \text{ years.}$$

34. If the voltmeter reads 6% too high, then the reading is 106% of the true voltage.
Let x = true voltage. Then, $106\%x = 120$
$$x = \dfrac{120}{106\%} = \dfrac{120}{1.06} = 113 \text{ volts (approximately)}$$

35. The error range is $\pm 0.15\%$(measurement) so
$$(\pm 0.15\%)(0.453) = (\pm.0015)(0.453) = 0.001$$
The upper boundary would be $0.453 + 0.001 = 0.454$
The lower boundary would be $0.453 - 0.001 = 0.452$

36. Let the short board = x and the longer one be *$3x + 1$*. Write the equation.
$$x + 3x + 1 = 21$$
$$4x + 1 = 21$$
$$4x + 1 - 1 = 21 - 1$$
$$4x = 20$$
$x = 5$ so the short board is 5 ft. and the longer is $3(5) + 1 = 16$ ft.

Chapter Two

Practice Set 2-1

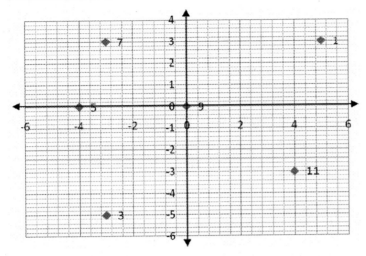

For problems 1-12, all answers are provided and plotted on one graph. Remember, for each (x,y) pair, only one dot appears on the graph. Always start at the origin, (0,0). The 1st number of the pair, the 'x', tells you how many units to move right, R, (if positive) or left, L, (if negative) and, the 2nd number, the 'y', tells you how far to move up, U, (if positive) or down, D, (if negative) from the location in which the first number was placed.

1. (5, 3) = 5 units right from the origin and then 3 units up
3. (-3, -5) = 3 L and 5 D
5. (-4, 0) = 4 L and none up or down (a point on the x-axis line)
7. (-3, 3) = 3 L and 3 U
9. (0, 0) = none R or L and none U or D (the origin)
11. (4, -3) = 4 R and 3 D

For problems 13-20, refer to the diagram on page 56 of the text. Always start at the origin and count units left, L, or right, R, first until you are lined up directly over or under the point, then count up, U, or down, D, until you reach the point. Also see figure 2-1 in the text for a quadrant map. Points on an axis line are not in a particular quadrant.

13. point F is 0 L or R and 0 U or D = (0, 0); i.e., it is at the origin
15. point A is 3 L and 4 U = (-3, 4); in quadrant II

17. point E is 3 L and 3 D = (-3, -3); in quadrant III
19. point G is 0 L or R and 3 D = (0, -3); on the y-axis line (i.e., not in a quadrant)

For numbers 21-29, refer to the diagram on page 56 of the text.
21. Only point C is on the positive x-axis, point F is at the origin where the x-coordinate = 0 and zero is not positive.
23. The y-coordinate is 0, (??,0) points F and C both match this requirement
25. "opposite of" means same absolute value but different signs; only point B has opposite numbers as its coordinates, (1,-1)
27. Is (3, 4) a solution for 2x - 5y = 26?
    Substitute: 2(3) - 5(4) = 6 - 20 = -14, not 26. Therefore, (3, 4) is not a solution.
29. Is (3, -2) a solution for 3x + 2y = 5?
    Substitute: 3(3) + 2(-2) = 9 + (-4) = 5. Therefore, (3, -2) is a solution.

For problems 30-40, we will find four solutions to complete the chart. First we let x = 0 and solve for y, 2nd we let x = 3 and solve for y, 3rd we let y = -2 and solve for x, and 4th we let y = 0 and solve for x.

	x = 0	x = 3	y = -2	y = 0
31. x + y = 8	0 + y = 8	3 + y = 8	x + (-2) = 8	x + 0 = 8
	y = 8	y = 5	x = 10	x = 8
	(0, 8)	(3, 5)	(10, -2)	(8, 0)
33. x - 3y = 12	0 - 3y = 12	3 - 3y = 12	x - 3(-2) = 12	x - 3(0) = 12
	-3y = 12	-3y = 9	x + 6 = 12	x - 0 = 12
	y = -4	y = -3	x = 6	x = 12
	(0, -4)	(3, -3)	(6, -2)	(12, 0)
35. x + 4y = 0	0 + 4y = 0	3 + 4y = 0	x + 4(-2) = 0	x + 4(0) = 0
	4y = 0	4y = -3	x - 8 = 0	x + 0 = 0

	$y = 0$	$y = -¾$	$x = 8$	$x = 0$
	(0, 0)	(3, -¾)	(8, -2)	(0, 0)

37. $y = x$

	$y = 0$	$y = 3$	$-2 = x$	$0 = x$
	(0, 0)	(3, 3)	(-2, -2)	(0, 0)

39. $-x - 2y = 6$

$-0 - 2y = 6$	$-3 - 2y = 6$	$-x - 2(-2) = 6$	$-x - 2(0) = 6$
$-2y = 6$	$-2y = 9$	$-x + 4 = 6$	$-x - 0 = 6$
$y = -3$	$y = -9/2 = -4.5$	$-x = 2$	$-x = 6$
		$x = -2$	$x = -6$
(0, -3)	(3, -4.5)	(-2, -2)	(-6, 0)

Practice Set 2-2

1. Let the y value = 0 in the equation and solve for *x*.

3. Calculate the coordinates of another point on the line by assigning either *x* or *y* a nonzero value and solving for the matching coordinate.

For problems 5-12 we did three different substitutions into each equation to complete the ordered pairs listed.

5. $x - y = 6$

$x - 0 = 6$	$0 - y = 6$	$x - (-2) = 6$
$x = 6$	$-y = 6$	$x + 2 = 6$
	$y = -6$	$x = 4$
(6, 0)	(0, -6)	(4, -2)

7. $2x + y = 6$

$2x + 0 = 6$	$2(0) + y = 6$	$2x + (-2) = 6$
$2x = 6$	$0 + y = 6$	$2x - 2 = 6$
$x = 3$	$y = 6$	$2x = 8$
		$x = 4$
(3, 0)	(0, 6)	(4, -2)

9. $y = -2$; This means that no matter what *x* may be chosen to be, *y* must be -2, so you may just write out the *x,y* pairs without any calculations: (3,-2), (0,-2), (any real number,-2)

11. $2x + 5y = 10$

$$\begin{array}{ccc} 2x + 5(0) = 10 & 2(0) + 5y = 10 & 2x + 5(-2) = 10 \\ 2x + 0 = 10 & 0 + 5y = 10 & 2x - 10 = 10 \\ 2x = 10 & 5y = 10 & 2x = 20 \\ x = 5 & y = 2 & x = 10 \\ (5, 0) & (0, 2) & (10, -2) \end{array}$$

For problems 13 - 40, follow Example 7 and the rule just above it in the text. In each case below, we chose to find the x-intercept first (let y = 0) and the y-intercept second (let x = 0).

13. $-x - 2y = 4$

$$\begin{array}{cc} -x - 2(0) = 4 & -0 - 2y = 4 \\ -x = 4 & -2y = 4 \\ x = -4 & y = -2 \\ (-4, 0) & (0, -2) \end{array}$$

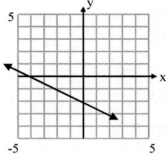

15. $-x - 2y = 5$

$$\begin{array}{cc} -x - 2(0) = 5 & -0 - 2y = 5 \\ -x - 0 = 5 & -2y = 5 \\ -x = 5 & y = -5/2 \\ x = -5 & \\ (-5, 0) & (0, -5/2) \end{array}$$

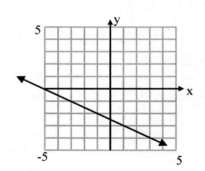

17. $2x + 3y = 6$

$$\begin{array}{cc} 2x + 3(0) = 6 & 2(0) + 3y = 6 \\ 2x = 6 & 3y = 6 \\ x = 3 & y = 2 \\ (3, 0) & (0, 2) \end{array}$$

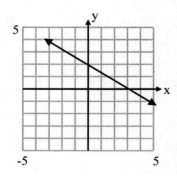

19. $3x - 5y = 15$

$\quad 3x - 5(0) = 15 \qquad 3(0) - 5y = 15$

$\qquad 3x = 15 \qquad\qquad -5y = 15$

$\qquad x = 5 \qquad\qquad\quad y = -3$

$\qquad (5, 0) \qquad\qquad\quad (0, -3)$

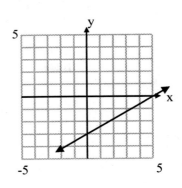

21. $4x - y = 3$

$\quad 4x - 0 = 3 \qquad 4(0) - y = 3$

$\qquad 4x = 3 \qquad\qquad -y = 3$

$\qquad x = ¾ \qquad\qquad y = -3$

$\qquad (¾, 0) \qquad\qquad (0, -3)$

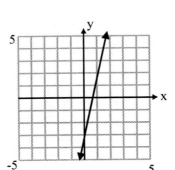

23. $3x = 2y$

$\quad 3x = 2(0) \qquad 3(0) = 2y$

$\qquad 3x = 0 \qquad\quad\; 0 = 2y$

$\qquad x = 0 \qquad\qquad 0 = y$

$\qquad (0, 0) \qquad\qquad (0, 0)$

Find another point by letting $y = 3$.

$\quad 3x = 2(3)$

$\quad 3x = 6$

$\quad x = 2 \quad (2, 3)$ Draw the line by connecting the origin to (2,3).

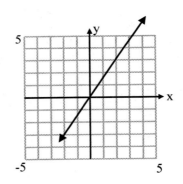

25. $x + y = 0$

$\quad x + 0 = 0 \qquad\qquad 0 + y = 0$
$\quad x = 0 \qquad\qquad\qquad y = 0$
$\quad (0, 0) \qquad\qquad\qquad (0, 0)$

Find another point by letting $y = 1$.

$\quad x + 1 = 0$

$\quad x = -1 \quad (-1, 1)$

Draw the line by connecting the origin to (-1, 1).

Fundamentals of Algebraic Modeling, 5th Ed.

27. $2x + 8 = -4y$; $2x + 4y = -8$

$2x + 4(0) = -8$ $2(0) + 4y = -8$
$2x = -8$ $4y = -8$
$x = -4$ $y = -2$
$(-4, 0)$ $(0, -2)$

29. $x = 2y + 2$; $x - 2y = 2$

$x - 2(0) = 2$ $0 - 2y = 2$
$x = 2$ $-2y = 2$
                           $y = -1$
$(2, 0)$ $(0, -1)$

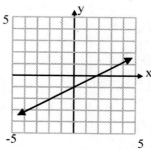

31. $y = 0.2x - 4$; $-0.2x + y = -4$

$-0.2x + 0 = -4$ $-0.2(0) + y = -4$
$-0.2x = -4$ $y = -4$
$x = 20$
$(20, 0)$ $(0, -4)$

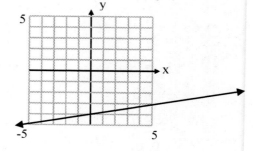

33. $x + ½y = 3$ $0 + ½y = 3$

$x + ½(0) = 3$ $½y = 3$
$x = 3$ $y = 6$
$(3, 0)$ $(0, 6)$

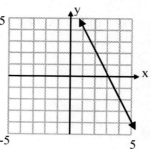

35. $x = -3$

     no place for $y = 0$ no y-intercept because
     so y can be any   x cannot $= 0$
     number, like 0
     $(-3, 0)$

Fundamentals of Algebraic Modeling, 5th Ed.

37. $3y = 2$; $y = \frac{2}{3}$

no x-intercept because y cannot = 0

so x may be any number you choose, like 0 $(0, \frac{2}{3})$

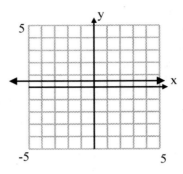

39. $3y + x = 4$

$\quad 3(0) + x = 4 \qquad 3y + 0 = 4$

$\qquad\qquad x = 4 \qquad\qquad 3y = 4$

$\qquad\qquad\qquad\qquad\qquad\quad y = \frac{4}{3}$

$\qquad (4, 0) \qquad\qquad (0, \frac{4}{3})$

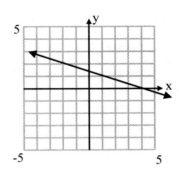

Practice Set 2-3

The slopes of the lines in problems 1-12 were all calculated following Examples 13, 14 and 15.

1. $m = \dfrac{1-(-4)}{6-8} = \dfrac{1+4}{-2} = -\dfrac{5}{2}$

3. $m = \dfrac{0-2}{-3-8} = \dfrac{-2}{-11} = \dfrac{2}{11}$

5. $m = \dfrac{-2-8}{-8-(-4)} = \dfrac{-10}{-4} = \dfrac{5}{2}$

7. $m = \dfrac{3-(-9)}{2.5-1} = \dfrac{12}{1.5} = 8$

9. $\dfrac{7}{1} = \dfrac{8-1}{x-4}$

$\quad 7(x - 4) = 1(7) \qquad$ [cross multiply]

$\quad 7x - 28 = 7$

$\qquad 7x = 35$

$\qquad\ x = 5$

Fundamentals of Algebraic Modeling, 5th Ed.

11. $\dfrac{7}{1} = \dfrac{-6-y}{3-7}$

   $7(-4) = 1(-6 - y)$ [cross multiply]

   $-28 = -6 - y$

   $-22 = -y$

   $22 = y$ [divide both sides by -1]

For numbers 13 - 25, the equations will be arranged in slope-intercept form, y = mx + b, where m is the slope of the line and b is the y-intercept (0,b). These graphs are approximations and you will find much more detailed graphs in the text.

13.   y = -4x + 1

   m = -4, y-intercept = (0, 1)

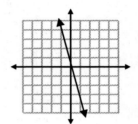

15.   3x + 2y = 6

   2y = -3x + 6

   $y = -\dfrac{3}{2}x + 3$

   $m = -\dfrac{3}{2}$, y-intercept = (0, 3)

17.   2x - 3y = 4

   -3y = -2x + 4

   $y = \dfrac{2}{3}x - \dfrac{4}{3}$

   $m = \dfrac{2}{3}$, y-intercept = $(0, -\dfrac{4}{3})$ or (0, -1.33)

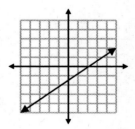

19. $4y - x = 10$

$4y = x + 10$

$y = \dfrac{1}{4}x + \dfrac{5}{2}$

$m = \dfrac{1}{4}$, y-intercept $= (0, \dfrac{5}{2}) = (0, 2.5)$

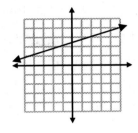

21. $\dfrac{3}{4}x - \dfrac{1}{2}y = \dfrac{5}{8}$

clear fractions, LCD = 8

$8\left(\dfrac{3}{4}x - \dfrac{1}{2}y\right) = 8\left(\dfrac{5}{8}\right)$

$6x - 4y = 5$

$-4y = -6x + 5$

$y = \dfrac{3}{2}x - \dfrac{5}{4}$

$m = \dfrac{3}{2}$, y-intercept $= (0, -\dfrac{5}{4}) = (0, -1.25)$

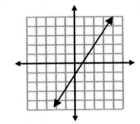

23. $y = -x$

$m = -1$, y-intercept $= (0, 0)$

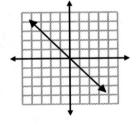

25. $4y = 12 + x$

$y = ¼x + 3$

$m = ¼$, y-intercept $= (0, 3)$

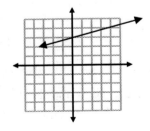

Fundamentals of Algebraic Modeling, 5th Ed.

For problems 26-29, first find the slope of the original line. This will be done by placing the equation given into slope-intercept form, y = mx + b, and reading the slope, m, directly from the equation. The slope of any line parallel to the original line will have the <u>same slope</u> as the original line's slope.

27. $3x - 8y = 11$

   $-8y = -3x + 11$

   $y = \frac{3}{8}x - \frac{11}{8}$; $m = \frac{3}{8}$

29. $-x + 6y = 2$

   $6y = x + 2$

   $y = \frac{1}{6}x + \frac{1}{3}$; $m = \frac{1}{6}$

For problems 31-35, first find the slope of the original line. This will be done by placing the equation given into slope-intercept form, y = mx + b, and reading the slope, m, directly from the equation. The slope of any line perpendicular to the original line will have a slope that is the <u>negative reciprocal</u> of the original line's slope.

31. $x + y = 3$

   $y = -x + 3$; $m = -1$; the slope of a perpendicular line = 1

33. $x = 2$; this is a vertical line and all vertical lines have an undefined slope ; any line perpendicular to a vertical line will be a horizontal line and all horizontal lines have a slope of zero; thus, the slope of a perpendicular line = 0

35. $y = 2$; $m = 0$; this is a horizontal line and any line perpendicular to it will be a vertical line; all vertical lines have undefined slopes; thus, the slope of a perpendicular line = undefined

For problems 36-40 we will find the slopes of both lines by placing both lines in slope-intercept form, y = mx + b, and then reading the slopes, m, directly from the equations. If the slopes are the same then the lines are parallel. If the slopes are negative reciprocals of each other then the lines are perpendicular. If neither of the above relationships exists, then the lines are neither parallel nor perpendicular.

37. x + y = 5   x - y = -5
    y = -x + 5   y = x + 5
    m = -1   m = 1

These two lines have slopes that are negative reciprocals of one another so they are perpendicular.

39. y = 2x   y = x
    m = 2   m = 1

These slopes are not the same nor are they negative reciprocals of each other so these lines are neither parallel nor perpendicular to each other.

Practice Set 2-4

1. Use the slope-intercept form y = mx + b and substitute: y = -2x + 8.

3. Use the slope-intercept form and substitute m = ½ and (x, y) = (6, -2) to find b.

    $-2 = ½(6) + b$

    $-2 = 3 + b$

    $-5 = b$

    Now substitute *m* and *b* into the slope intercept form to give y = ½x - 5.

5. Use the slope-intercept form and substitute m = -2 and (x, y) = (-8, -9) to find b.

    $-9 = -2(-8) + b$

    $-9 = 16 + b$

    $-25 = b$

    Now substitute *m* and *b* into the slope-intercept form to give y = -2x - 25.

7. Use the slope-intercept form and substitute m = $\frac{5}{2}$ and (x, y) = (-1, 4) to find b.

    $4 = \frac{5}{2}(-1) + b$

    $4 = -\frac{5}{2} + b$

    $4 + \frac{5}{2} = b$

    $\frac{8}{2} + \frac{5}{2} = b$

$$\frac{13}{2} = b$$

Now substitute *m* and *b* into the slope intercept form to give $y = \frac{5}{2}x + \frac{13}{2}$.

9. Use the slope-intercept form with $m = -\frac{1}{3}$ and the x-intercept point being (5, 0).

$$0 = -\frac{1}{3}(5) + b$$

$$0 = -\frac{5}{3} + b$$

$$\frac{5}{3} = b$$

Now substitute *m* and *b* into the slope-intercept form to give $y = -\frac{1}{3}x + \frac{5}{3}$.

For problems 11 - 22, the two points will be used to find the slope using the usual formula. Then the slope and one of the points (we used the first point given in each problem) will be used in the slope-intercept form of the equation of a line.

11. $m = \frac{9-3}{7-4} = \frac{6}{3} = 2$

   Now: $3 = 2(4) + b$

   $3 = 8 + b$

   $-5 = b$

   Substitute the values for *m* and *b*: $y = 2x - 5$

13. $m = \frac{-2-(-2)}{8-5} = \frac{0}{3} = 0$; a horizontal line

   Now: $-2 = 0(5) + b$

   $-2 = b$

   Substitute the values for *m* and *b*: $y = 0x - 2$ or $y = -2$

15. $m = \dfrac{2-4}{1-(-1)} = \dfrac{-2}{2} = -1$

   Now: $4 = -1(-1) + b$

   $4 = 1 + b$

   $3 = b$

   Substitute the values for *m* and *b*: $y = -x + 3$

17. $m = \dfrac{\frac{1}{3} - 1\frac{1}{3}}{5-7} = \dfrac{-1}{-2} = \dfrac{1}{2}$

   Now: $1\frac{1}{3} = \frac{1}{2}(7) + b$

   $1\frac{1}{3} = \frac{7}{2} + b$

   $\frac{4}{3} - \frac{7}{2} = \frac{8}{6} - \frac{21}{6} = -\frac{13}{6} = b$

   Substitute the values for *m* and *b*: $y = \dfrac{1}{2}x - \dfrac{13}{6}$

19. $m = \dfrac{-4.5 - 3}{-1 - (-4)} = \dfrac{-7.5}{3} = -2.5$

   Now: $3 = -2.5(-4) + b$

   $3 = 10 + b$

   $-7 = b$

   Substitute the values for *m* and *b*: $y = -2.5x - 7$

21. $m = \dfrac{1\frac{1}{3} - (-5)}{3 - 3} = \dfrac{6\frac{1}{3}}{0} = undefined$;

   a vertical line where x = constant = 3; in this case,   x = 3

23. A line with an undefined slope is a vertical line with x = constant. The given point of (1, 4) tells us that x = 1.

Problems 25-36 all relate to either parallel lines (lines having the same slopes) or perpendicular lines (lines having slopes that are negative reciprocals of one another). First find the slope by either rearranging the given equation or calculating it. If the lines are parallel, use the calculated slope. If the lines are perpendicular, find the negative reciprocal of the given line. Then use the slope-intercept form of the line to write the equation as you have done in previous problems.

Fundamentals of Algebraic Modeling, 5th Ed.

25. $2x + y = 6$

   in slope-intercept form: $y = -2x + 6$; $m = -2$; parallel line's slope = -2

   Now: $-7 = -2(1) + b$

   $\qquad -7 = -2 + b$

   $\qquad -5 = b$

   Substitute the values for *m* and *b*: $y = -2x - 5$.

27. $x - 3y = -4$

   in slope-intercept form: $y = \frac{1}{3}x + \frac{4}{3}$; $m = \frac{1}{3}$; perpendicular line's slope = -3

   Now: $10 = -3(3) + b$

   $\qquad 10 = -9 + b$

   $\qquad 19 = b$

   Substitute the values for *m* and *b*: $y = -3x + 19$

29. $m = \dfrac{11-8}{-1-2} = \dfrac{3}{-3} = -1$; the slope of parallel lines also = -1

   Now: $9 = -1(-2) + b$

   $\qquad 9 = 2 + b$

   $\qquad 7 = b$

   Substitute the values for *m* and *b*: $y = -x + 7$

31. $m = \dfrac{3-1}{1-(-4)} = \dfrac{2}{5}$; the slope of perpendicular lines = $-\dfrac{5}{2} = -2.5$

   Now: $2 = -2.5(-5) + b$

   $\qquad 2 = 12.5 + b$

   $\quad -10.5 = b$

   Substitute the values for *m* and *b*: $y = -2.5x - 10.5$

33. The line $x = 4$ is a vertical line and thus has an undefined slope. Any line parallel to it will also be a vertical line with $x =$ constant. The given point has $x = 3$, so this is the constant value of x. The equation is $x = 3$.

35. The line x = 4 is a vertical line and thus has an undefined slope. Any line perpendicular to it will be a horizontal line with y = constant. The given point has y = 3, so this is the constant value of y. The equation is y = 3.

37. The x-intercept = (3, 0) and the y-intercept = (0, -1). From these two points the slope may be found and then use the point-slope form of the equation of a line.

$$m = \frac{0-(-1)}{3-0} = \frac{1}{3}$$

Now: $0 = \frac{1}{3}(3) + b$

$0 = 1 + b$

$-1 = b$

Substitute the values for *m* and *b*: $y = \frac{1}{3}x - 1$

39. Passes through the origin (0, 0) and has m = -1.

Since the line passes through the origin, its y-intercept *(b)* value is 0.

Substitute the values for *m* and *b*: y = -1x + 0 or y = -x.

Practice Set 2-5

For problems 1 – 12, you will need to locate the y-intercept and determine the slope. Then substitute the y-intercept *(b)* and the slope *(m)* into the slope-intercept form for an equation of a line.

1. y-intercept = (0, 0) and the slope = $-\frac{1}{2}$

   Substitute the values for *m* and *b*: $y = -\frac{1}{2}x$

3. y-intercept = (0, 2) and the slope = -2

   Substitute the values for *m* and *b*: y = -2x + 2

5. y-intercept = (0, -3) and the slope = $\frac{1}{2}$

   Substitute the values for *m* and *b*: $y = \frac{1}{2}x - 3$

Fundamentals of Algebraic Modeling, 5th Ed.

7. y-intercept = (0, 2) and the slope = $\frac{1}{3}$

   Substitute the values for *m* and *b*: $y = \frac{1}{3}x + 2$

9. y-intercept = (0, 0) and the slope = 1

   Substitute the values for *m* and *b*: $y = x + 0$ or $y = x$

11. y-intercept = 3 and the slope = 0

    Substitute the values for *m* and *b*: $y = 0x + 3$ or $y = 3$

13a.   No

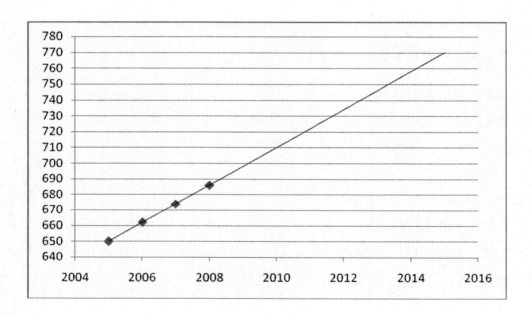

b.   Choose two points to calculate the slope.

(2005, 650) and (2006, 662)

$$m = \frac{662 - 650}{2006 - 2005} = \frac{12}{1} = 12 = \text{rate of change}$$

So rate of change = 12 students per year

c.   Since 2005 is the initial value, the y-intercept is 650.

Substitute the values of *m* and *b*: $y = 12x + 650$

d. Since the x = number of years after 2005, the x value is 8 for the year 2013.

Substitute x = 8 into the equation y = 12(8) + 650 = 96 + 650 = 746

The number of students enrolled in 2013 is predicted to be 746.

15. a.
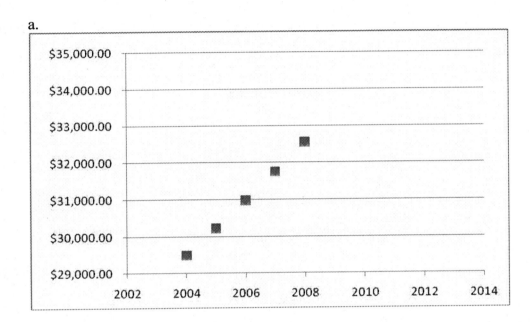

b. The rate of change is not constant so you cannot calculate a constant slope.
c. Since the rate of change is not constant, you cannot create a linear equation.
d. From the graph, it would appear that the salary will be more than $35,000. Since the graph is not linear, this is a general estimate.

17. Choose two points to find the slope or rate of change such as (2006, 119145) and (2007, 111892).

$$m = \frac{111892 - 119145}{2007 - 2006} = \frac{-7253}{1} = -7253 = \text{rate of change}$$

This means a reduction of 7253 tons per year is the rate of change.

No, because a constant rate for the next 7 years would be improbable since there is only so much refuse that can be recycled thus reducing the total amount.

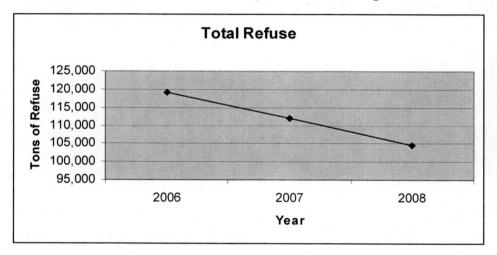

19. These rates do not have a linear trend since the changes in price are not consistent throughout the years.

Chapter 2 Review Problems

For problems 1 - 4, all points are to be plotted on one graph. Remember, for each (x,y) pair, only one dot appears on the graph. Always start at the origin, (0,0). The 1st number of the pair, the 'x', tells you how many units to move right, R, (if positive) or left, L, (if negative) and, the 2nd number, the 'y', tells you how far to move up, U, (if positive) or down, D, (if negative) from the location in which the first number was placed. See figure 2-1 in the text for a quadrant map. Points on an axis line are not in any quadrant.

1. (2,4) = 2 R and 4 U ; in quadrant I
2. (-3,4) = 3 L and 4 U ; in quadrant II
3. (-2,-1) = 2 L and 1 D ; in quadrant III
4. (0,6) = none R or L and 6 U ; on the y-axis

5. If you substitute 1 for x and -4 for y into the equation the result will be 6 if this pair of numbers is a solution: $2x - y = 6$

$2(1) - (-4) = 2 + 4 = 6$, so $(1, -4)$ is a solution.

In problems 6-10, substitute the value given in the (x, y) pair and solve for the other value.

6. $2x - y = 8$; $(-2, ?)$
$2(-2) - y = 8$
$-4 - y = 8$
$-y = 12$
$y = -12$; $(-2, -12)$

7. $y = -7x + 2$; $(6, ?)$
$y = -7(6) + 2$
$y = -42 + 2$
$y = -40$; $(6, -40)$

8. $3x + 2y = 8$; $(?, 4)$
$3x + 2(4) = 8$
$3x + 8 = 8$
$3x = 0$
$x = 0$; $(0, 4)$

9. $y = 7$; $(?, 7)$
This is a horizontal line and the values of x can be any real number.
(any real number, 7)

10. $5x - 3y = 11$; $(0, ?)$
$5(0) - 3y = 11$
$-3y = 11$
$y = -11/3$; $(0, -11/3)$

For problems 11-16, follow Example 7 and the rule just above it in the text. In each case below, we chose to find the x-intercept first (let y = 0) and the y-intercept second (let x = 0).

11. $y = \frac{1}{3}x - 1$; $\frac{1}{3}x - y = 1$

$\frac{1}{3}x - 0 = 1$ $\frac{1}{3}(0) - y = 1$

$\frac{1}{3}x = 1$ $-y = 1$

$x = 3$ $y = -1$

$(3, 0)$ $(0, -1)$

$m = \frac{1}{3}$

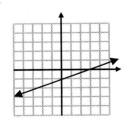

Fundamentals of Algebraic Modeling, 5th Ed.

12. x = -4

    no place for y = 0   this line is a vertical

    so y may be any number   line and thus has

    you choose, like 0   no y-intercept

    (-4, 0)

    m = undefined

13. 2x + y = 5

    2x + 0 = 5   2(0) + y = 5

    2x = 5   y = 5

    x = 2.5

    (2.5, 0)   (0, 5)

    m = -2

14. -3x + 2y = 6

    -3x + 2(0) = 6   -3(0) + 2y = 6

    -3x = 6   2y = 6

    x = -2   y = 3

    (-2, 0)   (0, 3)

    m = 3/2

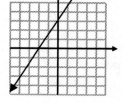

15. x - 3y = -3

    x - 3(0) = -3   0 - 3y = -3

    x = -3   -3y = -3

                      y = 1

    (-3, 0)   (0, 1)

    m = 1/3

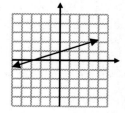

16. y = 3 is a horizontal line

no x-intercept because y cannot = 0

and x may be any number you choose, such as 0

which gives (0,3) as a point on the line (y-intercept)

m = 0

The slopes of the lines in problems 17-20 were all calculated following Examples 13, 14 and 15.

17. $m = \dfrac{4-7}{1-2} = \dfrac{-3}{-1} = 3$

18. $m = \dfrac{4-(-10)}{-2-5} = \dfrac{14}{-7} = -2$

19. $m = \dfrac{6-8}{-3-(-3)} = \dfrac{-2}{0} =$ undefined

20. $m = \dfrac{6-(-1)}{-4-(-6)} = \dfrac{6+1}{-4+6} = \dfrac{7}{2}$

In problems 21-25, the equation will first be written using the slope-intercept form.

21. Use the slope-intercept form: y = mx + b

$\qquad$ y = 2x + (-3) or 2x - y = 3

22. First find the slope of the line then use the point-slope form:

$m = \dfrac{6-(-4)}{-2-(-3)} = \dfrac{6+4}{-2+3} = \dfrac{10}{1} = 10$

Next use the slope-intercept form to find b.

6 = 10(-2) + b

6 = -20 + b

26 = b

Now substitute *m* and *b* into the slope-intercept form to give: y = 10x + 26 or

$\qquad\qquad\qquad\qquad\qquad$ 10x - y = -26 (in general form)

23. Use the slope-intercept form with the given slope and point to find b.

$$4 = -2(3) + b$$
$$4 = -6 + b$$
$$10 = b$$

Now substitute *m* and *b* into the slope-intercept form to give: $y = -2x + 10$ or

$$2x + y = 10 \text{ (in general form)}$$

24. Find the slope of the line given and the negative reciprocal of that line's slope will be the slope of the line we are looking for. Then use the slope-intercept form: $y = 2x - 5$ is written in slope-intercept form, so the slope is 2 and the negative reciprocal of 2 is $-½$.

$$5 = -½(-2) + b$$
$$5 = 1 + b$$
$$4 = b$$

Now substitute *m* and *b* into the slope-intercept form to give: $y = -½x + 4$.
Now multiply by 2 to clear the fraction if you want to write the equation in general form $x + 2y = 8$.

25. Parallel lines have the same slope so we first find the slope of the given line and then use the slope-intercept form:

$$3x - 4y = 9$$
$$-4y = -3x + 9$$
$$y = ¾x - 9/4 \text{ ; so the slope of the line is } ¾$$

Now substitute *m* and *b* into the slope intercept form to give: $y = ¾x + 4$
If you want to write the equation in general form, then rewrite as $¾x - y = -4$. Now multiply by 4 to clear the fraction to give $3x - 4y = -16$.

26. The y-intercept is (0, -1) and the slope is ½.

Substitute *m* and *b* into the slope-intercept form: $y = ½x - 1$

If you want to write the equation in general form then rewrite as $-½x + y = -1$. Now multiply by -2 to clear the fraction to give $x - 2y = 2$.

27. The y-intercept is (0, 4) and the slope is $-½$.

Substitute *m* and *b* into the slope-intercept form: $y = -½x + 4$

If you want to write the equation in general form, then rewrite as $½x + y = 4$. Now multiply by 2 to clear the fraction to give $x + 2y = 8$.

Chapter 2 Test

For problems 1-3, all points are to be plotted on one graph. Remember, for each (x, y) pair, only one dot appears on the graph. Always start at the origin, (0, 0). The 1st number of the pair, the 'x', tells you how many units to move right, R, (if positive) or left, L, (if negative) and, the 2nd number, the 'y', tells you how far to move up, U, (if positive) or down, D, (if negative) from the location in which the first number was placed. See figure 2-1 in the text for a quadrant map. Points on an axis line are not in any quadrant. See back of textbook for the graph.

1. (-2, 3) = 2 L and 3 U ; in quadrant II
2. (3, -1) = 3 R and 1 D ; in quadrant IV
3. (-3, -4) = 3 L and 4 D ; in quadrant III
4. If you substitute -3 for x and 5 for y into $3x + 2y$ is the result -1?

    $3(-3) + 2(5) = -9 + 10 = 1$; Since it is not -1, (-3, 5) is not a solution.

For problems 5-9 there are several ways to proceed. We placed the equations in slope-intercept form to find the slope and y-intercept, then solved for the x-intercept. To find the x-intercept let y = 0 and the y-intercept is read directly from the equation. The graphs use the intercepts.

5. $-4x + y = -4$
    $y = 4x - 4$
    $m = 4$
    y-intercept = (0, -4)
    solving for the x-intercept: $0 = 4x - 4$
    $-4x = -4$
    $x = 1$; (1, 0)

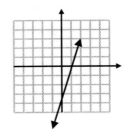

6. $2x + y = -4$
    $y = -2x - 4$
    $m = -2$
    y-intercept = (0, -4)
    solving for the x-intercept: $0 = -2x - 4$
    $2x = -4$
    $x = -2$; (-2, 0)

7. x = 5

   this is a vertical line
   m = undefined
   y-intercept = none
   x-intercept = (5, 0)

8. x + y = 3

       y = -x + 3
   m = -1
   y-intercept = (0, 3)
   solving for the x-intercept: 0 = -x + 3
                                           x = 3 ; (3, 0)

9. y = -2

   this is a horizontal line
   m = 0
   y-intercept = (0, -2)
   x-intercept = none

10. $m = \dfrac{y_2 - y_1}{x_2 - x_1} = \dfrac{-2-(-1)}{-3-2} = \dfrac{-1}{-5} = \dfrac{1}{5}$

11. A line perpendicular to a line with slope -3 would have slope $\dfrac{1}{3}$ since perpendicular lines have slopes that are negative reciprocals of each other.

12. All vertical lines have *undefined* slope.

13. In the line y = x, *m* (the coefficient of x) has a value of 1.

14. Use the slope-intercept form to find the value of b using the given information.

    6 = -3(-1) + b
    6 = 3 + b
    3 = b

    Now substitute *m* and *b* into the slope-intercept form to give: y = -3x + 3 or in general form: 3x + y = 3

15. Use the slope-intercept form to find the value of b using the given information.

    1 = ½ (-6) + b
    1 = -3 + b
    4 = b

    Now substitute *m* and *b* into the slope-intercept form to give: y = ½x + 4 or in general form: ½x - y = -4. Multiplying by the LCD gives x - 2y = -8.

Fundamentals of Algebraic Modeling, 5th Ed.

16. First find the slope.

$$m = \frac{-4-2}{0-1} = \frac{-6}{-1} = 6$$

Now use the slope-intercept form to find the value of b.

   -4 = 6(0) + b

   -4 = b

Now substitute *m* and *b* into the slope-intercept form to give: y = 6x - 4 or in general form: 6x - y = 4

17. First find the given line's slope. The slope of the perpendicular line we want will be the negative reciprocal of that slope. Then, use the slope-intercept form.

   2x - y = 3

   y = 2x - 3   m = 2 so the slope of our line will be -½

Now substitute *m* and *b* into the slope-intercept form to give: y = -½x + 4

To write in general form, reorder as ½x + y = 4 and multiply by 2 to clear the fraction which gives x + 2y = 8.

18. Parallel lines have the same slope so our line will have the same slope as the given line. Find the slope of the given line. x = 5 is a vertical line so its slope is undefined; thus, our line must also be vertical and can only have one value for x. In the given point, (2, -3), x is 2, so our line is x = 2.

19. The y-intercept = (0, 0) and the slope = -¾.
    Substitute *m* and *b* into the slope-intercept form of the equation.   y = -¾x
    To write in general form, rewrite as ¾x + y = 0 and then multiply by 4 to clear the fraction to give 3x + 4y = 0.

20. The y-intercept = (0, 0) and the slope = ¾.
    Substitute *m* and *b* into the slope-intercept form of the equation.   y = ¾x
    To write in general form, rewrite as -¾x + y = 0 and then multiply by -4 to clear the fraction to give 3x – 4y = 0.

Chapter Three

Practice Set 3-1

1. A function is a relation or rule in which, for each input value, there is exactly one output value. The independent variable is the input and is the variable that we control. The dependent variable is the output and its value is a result of the original choice of the value of the independent variable.

3. a) On April 23, the table gives a high temperature of 68°.
   b) The daily high temperature is a function of the date chosen. In other words, the high temperature depends on which date you choose. However, the date does not depend on the high temperature chosen because several dates could have the same high temperature.
   c) The dependent variable is the high temperature.
   d) The domain is the set of dates {20, 21, 22, 23, 24, 25, 26}.

5. a) The independent variable is the year (x-axis) and the dependent variable is the population (y-axis).
   b) The domain (values of x) are 1975-2005. The range (values of y) are 4.2 billion - 6.4 billion.

7. The grade, G, is a function of the identification number, N, since there is only one grade per student.

9. As the weight of a bag increases, the price of the bag increases. Therefore, the price is dependent on the weight of the potatoes.

11. As the slope of a hill increases, the speed of the scooter going uphill decreases. Therefore, the speed is dependent on the steepness of the hill.

13. As a person's age increases, his target heart rate when exercising decreases so the target heart rate when exercising is dependent on a person's age.

15. 9:45 pm

17. January 2 and December 30

19. 10:15 pm is the latest and 2:50 pm is the earliest

21.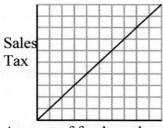
Sales Tax / Amount of food purchased

23.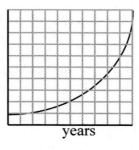
% of homes with a computer / years

25.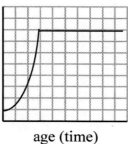
height / age (time)

27. d = 315 feet at a speed of 70 mph;

The graph is not a straight line. The faster the speed, the longer the stopping distance.

Stopping Distance (in ft) / Speed

29.

Interest Rate	5.5%	5.75%	6%	6.25%	6.5%	6.75%	7%
Payment	$709.74	$729.47	$749.44	$769.65	$790.09	$810.75	$831.63

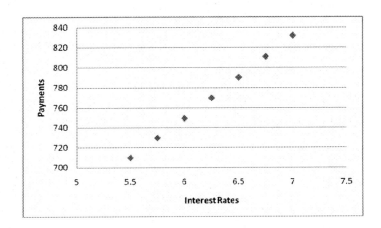

The graph looks similar to a straight line but is not perfectly linear.

Practice Set 3-2

1. $g(0) = 3(0) - 5 = -5$

3. $g(10) = 3(10) - 5 = 30 - 5 = 25$

5. $y = 2x + 1$

   $f(x) = 2x + 1$; Replace y with f(x)

   $f(-2) = 2(-2) + 1 = -3$

7. $3x^2 - y = 1$; Solve for y.

   $-y = -3x^2 + 1$ Divide by -1 and then replace y with f(x).

   $f(x) = 3x^2 - 1$

   $f(-2) = 3(-2)^2 - 1 = 3(4) - 1 = 12 - 1 = 11$

9. $5x - 2y = 8$; Solve for y.

   $-2y = -5x + 8$ Divide by -2 and replace y with f(x).

   $f(x) = \frac{5}{2}x - 4$

   $f(-2) = \frac{5}{2}(-2) - 4 = -5 - 4 = -9$

11. $f(2) = -16(2)^2 - 80(2) + 800 = -16(4) - 80(2) + 800 = -64 - 160 + 800 = 576$

   This number tells us that 2 sec after the ball is thrown, it is 576 ft above the ground.

13. $h(46.5) = 2.32(46.5) + 65.53 = 107.88 + 65.53 = 173.41$ cm

15. Since the equation uses cm, we must first convert his height from m to cm.

$$1.83 \text{ m} = 183 \text{ cm}$$

   Now we use this number in the function to solve for x (length of the femur).

$$h(x) = 2.32x + 65.53$$
$$183 \text{ cm} = 2.32x + 65.53$$
$$183 - 65.53 = 2.32x + 65.53 - 65.53$$
$$117.47 = 2.32x$$
$$50.63 \text{ cm} = x$$

17. a) $A(n) = 2^n$

   $A(30) = 2^{30} = 1,073,741,824$ pennies

   b) $1,073,741,824¢ = \$10,737,418.24$

19. a) (128 tickets)($7.50 per ticket) = $960

   b) $M(n) = \$7.50n$

   c) domain = 0 - 200 people; (any number of people from no one to full capacity may attend the movie)

   range = $0 - $1500; (amount of money earned from no one attending to full capacity)

21. Recommended heartbeats for a 20-year old would be f(20).

   $f(20) = -0.85(20) + 187 = 170$

   $f(60) = -0.85(60) + 187 = 136$; These two values of the function suggest that as a person gets older, his target heart rate decreases.

23. Reading the table for a value of 3 in the left column, f(3) = $29,080. This means that a person with 3 teaching years experience will make an annual salary of $29,080. The independent variable is the number of years of experience and the dependent variable is the salary that person earns based on experience.

25. f(4) = 0721 which means that the sun rises at 7:21am on January 4th; g(10) = 1654 which means that the sun sets on January 10th at 1654 hours or 4:54pm; the

independent variable is the date and the dependent variables are the sunrise and sunset times.

27. a) Locate the number 5 on the horizontal axis and then the y-value or the point at that location: f(5) = 75 mi.

   b) Using the graph, the value is approximately 130 miles (127.5 using the formula)

29. Locate the value 100 on the horizontal axis and find the y-value associated with that location: f(100) = $30.

   This means that it cost $30 to rent this car and drive 100 miles. The domain is numbers ≥ 0 (distances can't be negative) and the range is numbers $\geq$ $25 (minimum amount just to rent the car).

Practice Set 3-3

1. a)

x	f(x) = 1.75x	f(x)
1	1.75(1) =	$1.75
3	1.75(3) =	$5.25
5	1.75(5) =	$8.75
7	1.75(7) =	$12.25
9	1.75(9) =	$15.75

   b) initial value is 0; rate of change is $1.75 per mile

   c) f(x) = $1.75x where x = miles

   d) f(8) = 1.75(8) = 14 which means that an 8 mile taxi ride will cost $14.00

3. a) Based on the linear function model f(x) = initial value + (rate of change)(x) the initial value of this function is 5.5 million.

   b) The rate of change is 0.02 = 2% per year

   c) P(5) = 5.5 + 0.02(5) = 5.6 million

5. a) C(x) = $525 + $17.50x where x = number of guests

   b) C(200) = 525 + 17.50(200) = $4025

   c) We need to use the function to solve for x, the number of guests if the cost is $3150.

$$C(x) = 525 + 17.50(x)$$
$$3150 = 525 + 17.50x$$

Fundamentals of Algebraic Modeling, 5th Ed.

$$3150 - 525 = 525 - 525 + 17.50x$$
$$2625 = 17.50x$$
$$150 \text{ guests} = x$$

7. a) $\dfrac{1800-100}{5}$ = $340/year loss of value or rate of change of -$340 per year

   b) V(x) = initial value + (rate of change)(x) = 1800 - 340x where x = years

   c) V(4) = 1800 - 340(4) = 440 which means that after 4 years, the computer is worth only $440

9. First use two ordered pairs to find the slope, or rate of change.

$$m = \frac{y_2 - y_1}{x_2 - x_1} = \frac{-2-(-1)}{10-5} = \frac{-1}{5} = -\frac{1}{5}$$

Now use the slope-intercept form to find b and write the equation. Use the point (5, -1).

$$y = mx + b$$
$$-1 = -\frac{1}{5}(5) + b$$
$$-1 = -1 + b$$
$$0 = b$$
$$y = -\frac{1}{5}x + 0 \text{ or } f(x) = -\frac{1}{5}x$$

11. First use two ordered pairs to find the slope, or rate of change

$$m = \frac{5-5}{-1-2} = \frac{0}{-3} = 0$$

Now use the slope-intercept form to find *b* and write the equation. Use the point (2, 5).

$$y = mx + b$$
$$5 = 0(2) + b$$
$$5 = b$$
$$y = 0x + 5 \text{ or } f(x) = 5$$

13. a) rate of change = $\dfrac{650-450}{2006-1996} = \dfrac{200}{10} = +20$/year

   b) $f(x) = 450 + 20x$ where x = years after 1996

   c) $f(15) = 450 + 20(15) = 750$ which means that in the year 2011, the predicted population is 750 people.

15. a) rate of change = $\dfrac{1{,}000{,}000 - 600{,}000}{2008 - 2003} = \dfrac{400{,}000}{5} = \$80{,}000$/year

   b) $f(x) = \$600{,}000 + \$80{,}000x$ where x = number of years after 2003

   c) Use the function to solve for x, allowing the \$1,500,000 figure to be f(x).

   $f(x) = \$600{,}000 + \$80{,}000x$

   $\$1{,}500{,}000 = \$600{,}000 + \$80{,}000x$

   $\$1{,}500{,}000 - \$600{,}000 = \$600{,}000 - \$600{,}000 + \$80{,}000x$

   $\$900{,}000 = \$80{,}000x$

   $11.25 = x$

   That means the company's sales are predicted to reach \$1,500,000 in 11.25 years or during the year 2015.

17. a) rate of change = $\dfrac{4115 - 3536}{2006 - 2005} = \dfrac{579}{1} = \$579$/year

   b) $f(x) = \$3536 + \$579x$ where x = number of years after 2005

   c)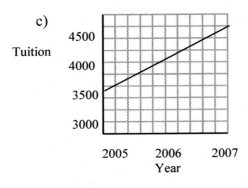

   d) $f(6) = 3536 + 579(6) = \$7010$

19. a) $f(x) = 100 + 5x$ where x = inches over 60 inches   or

   $f(x) = 100 + 5(x - 60)$ where x = woman's height in inches

   b) $f(7) = 100 + 5(7) = 135$ lbs

   $f(67) = 100 + 5(67 - 60) = 135$ lbs

Practice Set 3-4

1. direct - As the diameter increases, there is a corresponding increase in circumference. to check out.

3. inverse - As the amount of water coming in from the hose increases, the time it takes to fill the swimming pool decreases.

5. direct - As the amount of water used in a month increases, the water bill increases.

7. $y = kz$

9. $a = kbc$

11. $d = kef^3$

13. $m = \dfrac{k\sqrt{n}}{p^3}$

15. $y = kx$

$8 = k(24)$

$k = \dfrac{24}{8} = \dfrac{1}{3}$

So, $y = \left(\dfrac{1}{3}\right)x$ is the working formula

$y = \left(\dfrac{1}{3}\right)(36) = 12$

17. $y = \dfrac{k}{x}$

$9 = \dfrac{k}{6}$

$k = 54$

So, $y = \dfrac{54}{x}$ is the working formula

$y = \dfrac{54}{36} = 1.5$

19. $y = k\sqrt{x}$

$24 = k\sqrt{16}$

$24 = 4k$

$k = 6$

So, $y = 6\sqrt{x}$ is the working formula.

$y = 6\sqrt{36} = 6 \cdot 6 = 36$

21. $y = kxz^2$

$150 = k(3)(5^2)$

$150 = 75k$

$k = 2$

So, $y = 2xz^2$ is the working formula.

$y = 2(12)(8^2) = 1536$

23. $p = \dfrac{kq}{r^2}$

$40 = \dfrac{k(20)}{4^2}$

$k = 32$

So, $p = \dfrac{32q}{r^2}$ is the working formula.

$p = \dfrac{32(24)}{8^2} = 12$

25. $a = kr^2$

$78.5 = k(5)^2$

$78.5 = 25k$

$k = 3.14$

So, $a = 3.14r^2$ is the working formula.

$a = 3.14(7)^2 = 3.14(49) = 153.86 \text{ cm}^2$

27. $d = kt^2$

$27 = k(3)^2$

$27 = 9k$

$k = 3$

So, d = (3)(t²) is the working formula.

$$d = (3)(2)^2 = 12 \text{ ft}$$

29. $p = k\sqrt{l}$

$$0.5 = k\sqrt{16}$$

$$k = 0.125$$

So, $p = 0.125\sqrt{l}$ is the working formula.

$$0.75 = 0.125\sqrt{l} \qquad \text{[divide both sides by 0.125]}$$

$$\frac{0.75}{0.125} = \sqrt{l}$$

$$6 = \sqrt{l} \qquad \text{[square both sides to eliminate the square root]}$$

$$36 \text{ in} = l$$

31. $f = \dfrac{k}{d^2}$

$$20 = \frac{k}{(2)^2}$$

$$k = 80$$

So, $f = \dfrac{80}{d^2}$ is the working formula.

$$5 = \frac{80}{d^2}$$

$$5d^2 = 80$$

$$d^2 = \frac{80}{5}$$

$$d^2 = 16$$

$$d = 4 \text{ in} \qquad \text{[take the square root of both sides to eliminate the square]}$$

33. $w = \dfrac{k}{d^2}$

$$180 = \frac{k}{(4000)^2}$$

$$k = 2{,}880{,}000{,}000$$

So, $w = \dfrac{2{,}880{,}000{,}000}{d^2}$ is the working formula.

$$w = \dfrac{2880000000}{(4500)^2}$$

$$w = 142.2 \text{ lb}$$

35. $P = \dfrac{kV^2}{R}$

$$180w = \dfrac{k(90)^2}{45}$$

$$(180)(45) = k(90)^2 \quad \text{[cross multiply]}$$

$$\dfrac{180(45)}{8100} = 1 = k \quad \text{[divide both sides by 8100]}$$

So, $P = \dfrac{(1)(V^2)}{R}$

$$P = \dfrac{(1)(120)^2}{30} = 480 \; W$$

37. $e = ktl$

$$2.7 = k(225)(100)$$

$$2.7 = k(22500)$$

$$k = 0.00012$$

So, $e = (0.00012)tl$ is the working formula.

$$e = (0.00012)(150)(40) = 0.72 \text{ in}$$

39. a) direct variation

b) $\dfrac{4}{3}\pi$

c) $V(9) = \dfrac{4}{3}\pi(9)^3 = 972\pi^3 = 3053.6 \; in^3$

Fundamentals of Algebraic Modeling, 5th Ed.

Practice Set 3-5

Note: for problems 1 through 5, the graphs are shown as drawn using the graphing function on a graphing calculator and the roots (x-axis intercept points), if any, and maximum or minimum points are found using the CALC function of a graphing calculator as explained on page 129 in the text. The roots may also be found by factoring or using the quadratic formula. Your graphs should match those in the text.

1.   minimum = -4
       roots = -1 and –5

3.   minimum = -2.3
       roots = -1 and –4

5.   maximum = 5.3
       roots = 1.7 and -1

7. $f(x) = 3x^2$
       $f(-3) = 3(-3)^2 = 3(9) = 27$

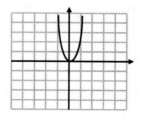

9. $f(x) = \dfrac{1}{3}x^3$

   $f(-3) = \dfrac{1}{3}(-3)^3 = \dfrac{1}{3}(-27) = -9$

11. $V = t^2 - 14t + 48$

   Substitute 3 in for V and solve for t.

   $3 = t^2 - 14t + 48$ (set = 0 and factor)

   $t^2 - 14t + 45 = 0$

   $(t - 9)(t - 5) = 0$

   $t = 9$ and $t = 5$

13. $0 = t^2 - 14t + 48$

   Factor and solve for t.

   $(t - 6)(t - 8) = 0$

   $t = 6$ and $t = 8$

15. When the rock hits the ground, the height is 0. Substitute 0 in for h and solve for t.

   $0 = 240 - 32t - 16t^2$

   $0 = -16(t^2 + 2t - 15)$

   $0 = -16(t - 3)(t + 5)$

   $t = 3$ and $t = -5$

   Since t represents time, t cannot equal -5. Therefore the rock will hit the ground in 3 seconds.

17. Use the CALC function on the graphing calculator to determine the maximum of this function. The x-coordinate represents the time, 0.2 s. The y-coordinate represents the height, 4.6 feet.

19. a) $f(2 \text{ sec}) = -16(2)^2 + 45(2) + 400 = -16(4) + 90 + 400 = -64 + 90 + 400 = 426$ feet

   b) Use the CALC function on the calculator to find the maximum y-value of the function: 431.64 ft.

   c) Look at the x-intercept on the graph to calculate when the ball reaches the ground. This happens when x = 6.6 seconds.

21. $p(1.5) = 920(1.5)^2 = \$2070$

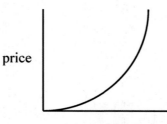

price

weight in carats

23. Substitute a = 1, b = -2 and c = -15 into quadratic formula and solve for x.

$$x = \frac{-(-2) \pm \sqrt{(-2)^2 - 4(1)(-15)}}{2(1)} = \frac{2 \pm \sqrt{4+60}}{2} = \frac{2 \pm \sqrt{64}}{2} = \frac{2 \pm 8}{2}$$

So x = 10/2 = 5 and x = -6/2 = -3

25. $3x^2 + 1 = 7x$

Set the equation equal to zero. $3x^2 - 7x + 1 = 0$

Substitute a = 3, b = -7 and c = 1 into quadratic formula and solve for x.

$$x = \frac{-(-7) \pm \sqrt{(-7)^2 - 4(3)(1)}}{2(3)} = \frac{7 \pm \sqrt{49-12}}{6} = \frac{7 \pm \sqrt{37}}{6}$$

so x = 2.18 and x = 0.15

27. $2x^2 - 3x = 9$

Set the equation equal to zero. $2x^2 - 3x - 9 = 0$

Substitute a = 2, b = -3, and c = -9 into the quadratic formula and solve for x.

$$x = \frac{-(-3) \pm \sqrt{(-3)^2 - 4(2)(-9)}}{2(2)} = \frac{3 \pm \sqrt{9+72}}{4} = \frac{3 \pm \sqrt{81}}{4} = \frac{3 \pm 9}{4}$$

so $x = \frac{12}{4} = 3$ and $\frac{-6}{4} = -1.5$.

29. $h(t) = 112t - 16t^2$

Substitute 160 in for the height and solve for t. $160 = 112t - 16t^2$

$16t^2 - 112t + 160 = 0$ (set equation equal to zero)

$16(t^2 - 7t + 10) = 0$

$16(t - 5)(t - 2) = 0$

t = 5 seconds and 2 seconds

Fundamentals of Algebraic Modeling, 5th Ed.

31. $h(t) = 560t - 16t^2$

   Substitute 1056 in for height and solve for t. $1056 = 560t - 16t^2$

   $16t^2 - 560t + 1056 = 0$ (set equation equal to zero)

   $16(t^2 - 35t + 66) = 0$

   $16(t - 33)(t - 2) = 0$

   t = 33 seconds and 2 seconds

   The rocket passes the balloon at t = 33 seconds on its way down.

33. $C = x^2 + 16x + 114$ (can also be done with the quadratic formula where a = 1, b = 16 and c = 114)

   Substitute 6450 in for the cost and solve for x.

   $6450 = x^2 + 16x + 114$

   $x^2 + 16x - 6336 = 0$ (set equal to zero and solve for x)

   $(x - 72)(x + 88) = 0$

   x = 72 and x = -88

   Since x represents the number of units manufactured, x cannot equal a negative number. Therefore x = 72 units.

35. $P(x) = 3x^2 + 240x - 1800$

   Substitute 70,200 in for the profit amount and solve for x.

   $70,200 = 3x^2 + 240x - 1800$

   $3x^2 + 240x - 72,000 = 0$ (set equal to zero and solve for x)

   $3(x^2 + 80x - 24,000) = 0$

   $3(x - 120)(x + 200) = 0$

   x = 120 and x = -200

   Since x represents the number of washing machines sold, x cannot equal a negative number.

   Therefore x = 120 washing machines.

37. a) Neither set of data is linear. The rate of change is not consistent.

   b) They are not linear.

Practice Set 3-6

1. $f(x) = 5^x$
   $f(2) = 5^2 = 25$

3. $f(x) = 3^x - 2$
   $f(2) = 3^2 - 2 = 9 - 2 = 7$

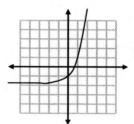

5. $f(x) = \left(\dfrac{1}{2}\right)^x$
   $f(2) = \left(\dfrac{1}{2}\right)^2 = \dfrac{1}{4}$

7. $f(x) = e^x$
   $f(2) = e^2 = 7.39$

9. $f(x) = e^{-0.5x}$
   $f(2) = e^{-0.5(2)} = e^{-1} = 0.3679$

11. $f(t) = 15{,}000(1.005)^t$

    Since t = 0 corresponds to the year 2005, substitute t = 10 for the year 2015.

    $f(10) = 15{,}000(1.005)^{10} = 15{,}000(1.051140132) = 15767.10198 \approx 15{,}767$ people

13. $A = 15{,}000 e^{rt}$

    $A = 15{,}000\, e^{(0.065)(10)} = 15{,}000\, e^{0.65} = 15{,}000(1.915540829) = 28733.112 \approx 28733$ people

15. $A = 12{,}500\, e^{(-0.008)(7)} = 12{,}500\, e^{-0.056} = 12{,}500(0.9455391359) = 11819.2392 \approx$ 11819 deer

17. $A = 12{,}576{,}800\, e^{(-0.0075)(10)} = 12{,}576{,}800\, e^{-0.075} = 12{,}576{,}800(0.9277434863) =$ $11{,}668{,}044.28 \approx 11{,}668{,}044$ people

    $A = 12{,}576{,}800\, e^{(-0.0075)(50)} = 12{,}576{,}800\, e^{-0.375} = 12{,}576{,}800(0.6872892788) =$ $8643899.801 \approx 8643900$ people

19. $A = 57.6\, e^{(-0.003)(17)} = 57.6\, e^{-0.051} = 57.6(0.9502786705) =$ $54.73605142 \approx 54.7$ million people

21. $y = y_0 e^{-0.04t}$

    $y = 155\, e^{(-0.04)(30)} = 155\, e^{-1.2} = 155(0.3011942119) = 46.68510285 \approx 46.69$ grams

    $y = 155\, e^{(-0.04)(120)} = 155\, e^{-4.8} = 155(0.008229747) = 1.275610793 \approx 1.28$ grams

23. $y = 300\, e^{-.032t}$

    $y = 300\, e^{(-0.032)(10)} = 300\, e^{-0.32} = 300(0.7261490371) = 217.8447111 \approx 217.84$ grams

25. $y = A e^{rt}$

    $y = 25{,}000\, e^{(0.045)(5)} = 25{,}000\, e^{0.225} = 25{,}000(1.252322716) = 31308.0679 =$ $\$31{,}308.07$

27. $y = 2\, e^{(0.035)(5)} = 2\, e^{0.175} = 2(1.191246217) = 2.382492433 = \$2.38$

    Draw the function $y = 2\, e^{0.035x}$ using your calculator and look at the table values. The value for x when y is more than 3 is 11.58 years.

Chapter 3 Review Problems

1. T($12) = $0.72 means that you pay tax of 72 cents on a purchase of $12.
2. Answers will vary but must fit the textbook definitions.
3. Both the domain and range must be greater than or equal to 0 because purchases and taxes may be positive amounts only.
4. As the age of a person increases, his height increases up to a certain age. Then it remains steady with some shrinkage in later years.
5. As time passes, the number of bacteria in a culture increases.

6. The faster the speed of a car, the less time it takes to drive from home to school (unless you get a speeding ticket!)

7. independent variable: number of minutes; dependent variable: cost

8. domain: ≥ 0 minutes   range: $\geq \$5.00$

9. $d = \dfrac{x^2}{20} + x = \dfrac{20^2}{20} + 20 = \dfrac{400}{20} + 20 = 20 + 20 = 40\,ft$

$d = \dfrac{x^2}{20} + x = \dfrac{40^2}{20} + 40 = \dfrac{1600}{20} + 40 = 80 + 40 = 120\,ft$

$d = \dfrac{x^2}{20} + x = \dfrac{60^2}{20} + 60 = \dfrac{3600}{20} + 60 = 180 + 60 = 240\,ft$

$d = \dfrac{x^2}{20} + x = \dfrac{80^2}{20} + 80 = \dfrac{6400}{20} + 80 = 320 + 80 = 400\,ft$

The graph is nonlinear with stopping distances increasing quickly.

10. $f(x) = 2x - 6$

   (a) $f(-2) = 2(-2) - 6 = -10$

   (b) $f(0) = 2(0) - 6 = -6$

   (c) $f(4) = 2(4) - 6 = 2$

   (d) $f(½) = 2(½) - 6 = -5$

NOTE: for problems 11 through 15, solve for y and then replace y with f(x).

11. $y = -3x + 2$ is already solved for y so, $f(x) = -3x + 2$

12. $-3y = 2x + 6$

   $y = -⅔x - 2$ (divide by -3)

   $f(x) = -⅔x - 2$

13. $x - y = 8$

   $-y = -x + 8$ (divide by -1)

   $y = x - 8$

   $f(x) = x - 8$

14. $3x + 2y = 6$

   $2y = -3x + 6$

   $y = -\frac{3}{2}x + 3$ (divide by 2)

   $f(x) = -\frac{3}{2}x + 3$

15. $4x = -y + 1$

   $y = -4x + 1$

   $f(x) = -4x + 1$

16. $f(x) = 2x^2 - 1$ and to fill in the chart, we will let $x = -2, -1, 0, 1,$ and 2.

   $f(-2) = 2(-2)^2 - 1 = 2(4) - 1 = 8 - 1 = 7$

   $f(-1) = 2(-1)^2 - 1 = 2(1) - 1 = 2 - 1 = 1$

   $f(0) = 2(0)^2 - 1 = 0 - 1 = -1$

   $f(1) = 2(1)^2 - 1 = 2(1) - 1 = 2 - 1 = 1$

   $f(2) = 2(2)^2 - 1 = 2(4) - 1 = 8 - 1 = 7$

17. Answers to a), b), and c) are found by inspection of the graph:

   a) $g(0) = -3$ b) $g(3) = 3$ c) $g(-2) = 0$

18. $C(120) = \$29 + \$0.06(120) = \$36.20$, this is the cost of renting the car and driving it 120 miles.

19. a) The base fee is \$75 plus \$30 times the number of credit hours, x, taken.

   $C(x) = \$75 + \$30x$

   b) $C(12) = \$75 + \$30(12) = \$75 + \$360 = \$435$

20. a) $f(x) = 1000 + 5.50x$ where x = number of books printed

   b) The domain is the set of natural numbers ≥ 250.

   c) $f(375) = 1000 + 5.50(375) = \$3062.50$

21. a) $f(x) = 1400 + 0.15(x - 14000)$ where x is taxable income

   b) $f(32,355) = 1400 + 0.15(32,355 - 14,000) = \$4153.25$

22. a) The base salary is $1000 plus 25% (0.25) times her dollars worth of sales, s

   $f(s) = \$1000 + 0.25s$

   b) $f(\$2500) = \$1000 + 0.25(\$2500) = \$1625$

23. Inspection of the table shows that when x increases by one, y increases by 2. The slope of the line is the change in y divided by the change in x, so slope = 2/1 = 2. Now use this slope and one of the x, y pairs, (2,1), from the table to find the equation of the line.

$$y = mx + b$$
$$1 = 2(2) + b$$
$$1 = 4 + b$$
$$-3 = b$$
$$f(x) = 2x - 3$$

24. Inspection of the table shows that when x increases by four, y increases by 2. The slope of the line is the change in y divided by the change in x so, the slope is $\frac{2}{4} = \frac{1}{2}$. Now use this slope and one of the x, y pairs, (0, 4), from the table to find the equation of the line.

$$y = mx + b$$
$$4 = \tfrac{1}{2}(0) + b$$
$$4 = b$$
$$f(x) = \tfrac{1}{2}(x) + 4$$

25. independent variable: time; dependent variable: length of fingernails. This is a direct variation because as times increases, the length of fingernails increases.

26. independent variable: temperature; dependent variable: time for snowman to melt. This is an inverse variation because as temperature increases, the time it takes for the snowman to melt decreases.

27. $s = kt$

28. $z = \dfrac{k}{p}$

29. $m = krs$

30. $x = \dfrac{k\sqrt[3]{z}}{y^2}$

31. $u = kv^3$

$8 = k(1)^3$

k = 8, so our working formula is: $u = 8v^3$

$u = 8(2)^3 = 8(8) = 64$

32. $x = \dfrac{k}{y^2}$

$2 = \dfrac{k}{3^2} = \dfrac{k}{9}$

k = 2(9) = 18, so our working formula is: $x = \dfrac{18}{y^2}$

$x = \dfrac{18}{5^2} = \dfrac{18}{25}$

x = 0.72

33. $s = ktg$

10 = k(3)(2.5)

10 = 7.5k

1.33 = k   so our working formula is: *s = 1.33tg*

s = 1.33(7)(6) = 55.86 = 56

34. $N = \dfrac{kL^2}{M^3}$

$9 = \dfrac{k(2)^2}{1^3}$

9 = 4k

2.25 = k   so our working formula is $N = \dfrac{2.25L^2}{M^3}$

$N = \dfrac{2.25(4)^2}{2^3} = \dfrac{2.25(16)}{8} = 2.25(2) = 4.5$

Fundamentals of Algebraic Modeling, 5th Ed.

35. $d = kt^2$

$64.4 = k(2)^2$

$64.4 = 4k$

$16.1 = k$ so our working formula is: $d = 16.1t^2$

$d = 16.1(3)^2 = 16.1(9) = 144.9$ feet

36. $f = \dfrac{k}{d^2}$

$50 = \dfrac{k}{12^2}$

$50(144) = k$

$7200 = k$ so our working formula is: $f = \dfrac{7200}{d^2}$

$f = \dfrac{7200}{20^2} = \dfrac{7200}{400} = 18$ units

37. The graphs for a) and b) were done by graphing calculator and, the roots and max/mins were found with the CALC feature following the in the text.

   a) minimum = -8; roots = -1, 3

b) maximum = 9; roots = -4, 2

38. Graphs should be done with the graphing calculator.

   (a) f(-3) = ½(-3)⁴ = ½(81) = 40.5

   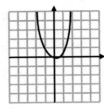

   (b) g(-3) = -2(-3)³ = -2(-27) = 54

   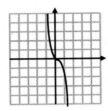

39. $x^2 - 10x - 4 = 0$

   Remember the quadratic formula is: $x = \dfrac{-b \pm \sqrt{b^2 - 4ac}}{2a}$

   Substitute a = 1, b = -10, and c = -4 into the formula and solve for x.

   $x = \dfrac{-(-10) \pm \sqrt{(-10)^2 - 4(1)(-4)}}{2(1)} = \dfrac{10 \pm \sqrt{100 + 16}}{2} = \dfrac{10 \pm \sqrt{116}}{2}$

   x = 10.385 and -.385

40. $s = 5t + 16t^2$

   Substitute 74 in for s and solve for t. $74 = 5t + 16t^2$

   Set equal to zero and solve for x.

   $0 = 16t^2 + 5t - 74$

   $0 = (16t + 37)(t - 2)$

   t = -37/16 and t = 2

   Since t represents time, t cannot equal a negative number.

   Therefore t = 2 seconds.

41. $p = -x^2 + 16x - 24$

   Substitute 40 in for p and solve for x. $40 = -x^2 + 16x - 24$

Set equal to zero and solve for x.

$x^2 - 16x + 64 = 0$

$(x - 8)(x - 8) = 0 \qquad x = 8$ telephones

42. Graphs should be done with the graphing calculator.

a) $f(-2) = 3^{-2} = \dfrac{1}{3^2} = \dfrac{1}{9}$

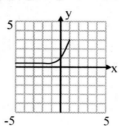

b) $g(-2) = (½)^{-2} = 2^2 = 4$

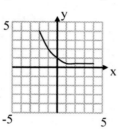

43. $P(10) = 100.35(1.0153)^{10} = 116.8$ million persons

44. $y = 25{,}000e^{(0.075)(24)} = 25{,}000\,e^{1.8} = 25{,}000(6.049647464) = 151241.1866$ or $151{,}242$ bacteria

45. $y = 500{,}000\,e^{(-0.05)(48)} = 500{,}000\,e^{-2.4} = 500{,}000(0.0907179533) = 45{,}358.9766$ or $45{,}359$ bacteria

46. $y = 850{,}000\,e^{(-0.02)(5)} = 850{,}000\,e^{-.1} = 850{,}000(0.904837418) = 769{,}111.8053$ or $769{,}112$ people

Chapter 3 Test

1. A linear function has a constant rate of change (slope).
2. $25,000 was the initial value of the car (purchase price) and it decreases in value by $1500 per year after it is purchased.
3. The cost of a house is a function of its square footage. The larger the amount of square footage, the more a house usually costs. The independent variable is the square footage and the dependent variable is the cost of the house.

4. a) The independent variable is time and the dependent variable is the number of widgets produced.

   b) The initial value is 0. The function is not perfectly linear.

NOTE: For problems 5 through 7, first solve for y and then replace y with f(x). Lastly find f(-4).

5. $2x - y = 4$

   $-y = -2x + 4$

   $y = 2x - 4$

   $f(x) = 2x - 4$

   $f(-4) = 2(-4) - 4 = -8 - 4 = -12$

6. $y = 5$

   $f(x) = 5$

   $f(-4) = 5$, constant

7. $3x^2 + y = -2$

   $y = -3x^2 - 2$

   $f(x) = -3x^2 - 2$

   $f(-4) = -3(-4)^2 - 2 = -3(16) - 2 = -50$

8. The initial value is $22,500 and it loses value (depreciates) at the rate of $2200 each year. $V(t) = \$22{,}500 - \$2200t$

9. $C(12) = 1.50 + 0.75(12) = 1.50 + 9.00 = \$10.50$, which means that a cab ride of 12 miles will cost $10.50.

10. a) $f(x) = \$21 + \$28x$ where x represents the number of credit hours taken

    b) domain is the set of whole numbers 12-15 and range is the set of numbers $357, $385, $413, $441

    c) $f(14) = 21 + 28(14) = \$413$ which means that a student who registers for 14 credit hours will pay $413

11. First, find the slope by choosing two ordered pairs from the table. We chose (-1,5) and (2, -4), to put in the slope formula: $m = \dfrac{5-(-4)}{-1-2} = \dfrac{9}{-3} = -3$. Now that we have the slope, we choose one point, (-1, 5) and find the equation using y = mx + b.

Fundamentals of Algebraic Modeling, 5th Ed.

$$5 = -3(-1) + b$$
$$5 = 3 + b$$
$$2 = b$$
$$y = -3x + 2$$
$$f(x) = -3x + 2$$

12. independent variable is time; dependent variable is the speed; The time it takes to finish a race decreases as the speed of the bike increases so this is inverse variation.

13. independent variable is time; dependent variable is the diameter of the tree; As time passes, the diameter of the trees increases so this is a direct variation.

14. $t = \dfrac{kB}{P^2}$

$7.5 = \dfrac{k(6)}{2^2} = \dfrac{6k}{4}$

$7.5(4) = 6k$

$\dfrac{30}{6} = k$

$5 = k$ so, our working formula is: $t = \dfrac{5B}{P^2}$

$t = \dfrac{5(63)}{3^2} = \dfrac{315}{9} = 35$

15. Her earnings, E, are directly proportional to the hours, H, worked. E = kH
    $260.50 = k(8)
    $32.5625 = k is her hourly pay rate, and our working formula is: E = $32.5625H
    E = $32.5625(35) = $1139.6875 = $1139.69

16. Let M = the coin's mass, T its thickness and R radius; $M = kTR^2$
    $26.4 = k(0.2)(2)^2$
    $26.4 = 0.8k$
    $\dfrac{26.4}{0.8} = k$
    $33 = k$ so our working formula is: $M = 33TR^2$
    $M = 33(0.3)(4)^2 = 158.4$ grams

17. Use a graphing calculator and the CALC function as shown in the text.

maximum = 9; roots = 3, -3

18. The graph was obtained with a graphing calculator.

$f(4) = -2^4 = -2 \cdot 2 \cdot 2 \cdot 2 = -16$

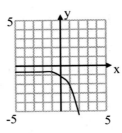

19. $C(18) = \$5200(1.045)^{18} = \$11,484.08959 = \$11,484.09$

20. $y = 260\, e^{(0.01)(50)} = 260\, e^{0.5} = 260(1.648721271) = 428.6675304 \approx 428.7$ million

21. $3x^2 - 11x = 7$ Set equal to zero and substitute into the quadratic formula.

$3x^2 - 11x - 7 = 0$ $a = 3, b = -11$ and $c = -7$

$$x = \frac{-(-11) \pm \sqrt{(-11)^2 - 4(3)(-7)}}{2(3)} = \frac{11 \pm \sqrt{121 + 84}}{6} = \frac{11 \pm \sqrt{205}}{6}$$

$x = 4.22$ and $x = -0.55$

22. $h = -16t^2 + 96t + 28$

a)

b) $-16(2)^2 + 96(2) + 28 = -16(4) + 96(2) + 28 = -64 + 192 + 28 = 156$ ft

c) $0 = -16t^2 + 96t + 28$

$a = -16, b = 96$ and $c = 28$

$$t = \frac{-96 \pm \sqrt{(96)^2 - 4(-16)(28)}}{2(-16)} = \frac{-96 \pm \sqrt{9216 + 1792}}{-32} = \frac{-96 \pm \sqrt{11008}}{-32}$$

$t = -.2787$ and $t = 6.279$ Since t represents time, t cannot equal a negative number. Therefore $t = 6.279$ seconds.

23. $y = 125{,}000\, e^{(0.03)(3)} = 125{,}000\, e^{0.09} = 125{,}000(1.094174284) =$
    $136771.7855 \approx 136{,}772$ people

Chapter Four

Practice Set 4-1

1. a) $(0.075)(\$25.55)=\$1.92$

   b) $(0.075)(\$155.75)=\$11.68$

   c) $(0.075)(\$256.18)=\$19.21$

3. state tax $= (0.0625)(\$25,615)=\$1600.94$

   County tax $= (0.02)(\$25,615)=\$512.30$

5. $x + (8\%)(x) = \$27,129.60$

   $1.08x = \$27,129.60$

   $x = (\$27,129.60)/(1.08) = \$25,120$ in sales

   sales tax $= \$27,129.60 - \$25,120 = \$2009.60$ or

   $(8\%)(\$25,120)=(0.08)(\$25,120)=\$2009.60$

7. $x + (7\%)(x) = \$213.95$

   $1.07x = \$213.95$

   $x = (\$213.95)/(1.07) = \$199.95$

9. Annual depreciation $= \dfrac{2150-500}{5} = \dfrac{1650}{5} = \$330$ per year.

11. Annual depreciation $= \dfrac{9450-1000}{7} = \dfrac{8450}{7} = \$1207.14$ per year.

13. $\$13,500 - \$11,880 = \$1620$ (amount depreciated)

    $\$1620/\$13,500 = 0.12 = 12\%$ depreciation

15. No pens made or sold but, fixed costs go on; a loss of $53,500

17. $(200,000)(\$1.05) - (200,000)(\$0.75) - \$53,500 =$ a profit of $6500

19. No games made or sold but, fixed costs continue; a loss of $8000

21. $(200,000)(\$5.00) - (200,000)(\$1.95) - \$8000 =$ a profit of $602,000

23. $(200)(6.50) + (200)(8.95) + (200)(13.5) - (200)(2.55) - (200)(3.05) - (200)(4.15) - 3500 =$ a profit of $340

25. (a) $C(x)=0.01x + 3500$

    (b) $C(20,500) = 0.01(20,500) + 3500 = \$3705$

    (c) $(20,500)(0.12) - 3705 =$ a loss of $1245

27. selling price $= \$35 + 55\%(\$35) = \$35 + 0.55(\$35) = \$54.25$

29. $48.51 - $26.95 = $21.56 markup

$21.56/$26.95 = 0.8 = 80% markup

Practice Set 4-2

1. $I = Prt$ and $M = P + I$

    $I = (500)(9\%)(1) = \$45$ interest; $M = 500 + 45 = \$545$

3. $I = Prt$ and $M = P + I$

    $I = (1200)(12\%)(1.5) = \$216.00$; $M = 1200 + 216 = \$1416.00$

5. $r = \dfrac{I}{Pt} = \dfrac{120}{(500)(2)} = 0.12 = 12\%$

7. $I = Prt = (\$250)(0.18)\left(\dfrac{3}{12}\right) = \$11.25$

9. $I = Prt$ and $M = P+I$

    $I = \$1200(12\%)(0.5) = \$72$; Maturity Value $= 1200 + 72 = \$1272$

    payment $= \$1272/6$ months $= \$212$ per month

11. $I = Prt$ and $M = P+I$

    $I = \$4500(10\%)(3) = \$1350$; Maturity Value $= 4500 + 1350 = \$5850$

    payment $= \$5850/36$ months $= \$162.50$ per month

13. a) Interest = total amount paid - cash price

    $I = \$2020 - \$1600 = \$420.00$

    b) Cost $= 24(\$80) + \$100 = \$2020.00$

    c) $r = \dfrac{I}{Pt} = \dfrac{420}{(1500)(2)} = 0.14 = 14\%$ (remember that only $1500 was financed.)

15. 10 payments $\times \$108.50 = \$1085.00$ total paid

    $\$1085.00 - \$1000 = \$85.00$ interest paid by financing

    $r = \dfrac{I}{Pt} = \dfrac{85.00}{(1000)\left(\dfrac{10}{12}\right)} = 0.102 = 10.2\%$

17. $M = P\left(1+\dfrac{r}{12}\right)^{12t} = 2000\left(1+\dfrac{0.072}{12}\right)^{12} = \$2148.85$

19. $M = P\left(1+\dfrac{r}{4}\right)^{4t} = 25,000\left(1+\dfrac{0.06}{4}\right)^{20} = \$33,671.38$

21. a) $M = P\left(1+\dfrac{r}{4}\right)^{4t} = 900\left(1+\dfrac{0.10}{4}\right)^{12} = \$1210.40$

   Interest = Maturity Value - Principal = $1210.40 - $900 = $310.40

   b) $M = P\left(1+\dfrac{r}{12}\right)^{12t} = 900\left(1+\dfrac{0.10}{12}\right)^{36} = \$1213.36$

   Interest = Maturity Value - Principal = $1213.36 - $900 = $313.36

23. $M = P\left(1+\dfrac{r}{12}\right)^{12t} = 5000\left(1+\dfrac{0.06}{12}\right)^{60} = \$6744.25$;   She will have more than $6500.

25. Monthly $= 10000\left(1+\dfrac{0.06}{12}\right)^{120} = \$18,193.97$

   Quarterly $= 10000\left(1+\dfrac{0.06}{4}\right)^{40} = \$18,140.18$

   Monthly - Quarterly = $53.79

27. $A = Pe^{rt}$
   $A = \$25{,}000 e^{(0.045)(5)} = 25000 e^{0.225} = \$31{,}308.07$

29. $M = Pe^{rt} = 140{,}000 e^{(0.02)(10)} = 140{,}000 e^{0.2} = \$170{,}996.39$

Practice Set 4-3

1. Unpaid Balance × Monthly Interest Rate = Interest Due

   $\$350.75 \times \dfrac{0.18}{12} = 350.75 \times 0.015 = \$5.26$

3. Unpaid Balance × Monthly Interest Rate = Interest Due

   $\$655.90 \times \dfrac{0.198}{12} = 655.90 \times 0.0165 = \$10.82$

5. Monthly Interest Rate $= \dfrac{0.18}{12} = 0.015$

   June Finance Charge: $350.10 × 0.015 = $5.25
   June Unpaid Balance: $350.10 + 5.25 + 25.00 + 36.75 - 75.00 = $342.10
   July Finance Charge: $342.10 × 0.015 = $5.13
   July Unpaid Balance: $342.10 + 5.13 + 150.40 - 200.00 = $297.63
   August Finance Charge: $297.63 × 0.015 = $4.46
   August Unpaid Balance: $297.63 + 4.46 + 208.75 + 55.40 - 150.00 = $416.24

7. Monthly Interest Rate = $\frac{0.18}{12} = 0.015$

   October Finance Charge: $59.75 \times 0.015 = \$0.90$

   October Unpaid Balance: $\$59.75 + 0.90 + 325.88 + 34.98 - 50.00 = \$371.51$

   November Finance Charge: $\$371.51 \times 0.015 = \$5.57$

   November Unpaid Balance: $\$371.51 + 5.57 + 59.06 + 295.60 - 150.00 = \$581.74$

   December Finance Charge: $\$581.74 \times 0.015 = \$8.73$

   December Unpaid Balance: $\$581.74 + 8.73 + 155.67 + 230.75 - 50.00 = \$926.89$

9. Monthly Interest Rate = $\frac{0.18}{12} = 0.015$

   June Fin. Chg.: $1295(0.015) = \$19.43$

   June Unpaid Bal.: $1295 + 19.43 + 65 - 120 = \$1259.43$

   July Fin. Chg.: $1259.43(0.015) = \$18.89$

   July Unpaid Bal.: $1259.43 + 18.89 + 1325 + 153.85 - 350 = \$2407.17$

   August Fin. Chg.: $2407.17(0.015) = \$36.11$

   August Unpaid Bal.: $2407.17 + 36.11 + 415.98 + 65.09 - 575 = \$2349.35$

11.

Date	Balance Due	Number of Days until Balance Changed	(balance due) × (number of days)
June 3	$145.90	14	14 × $145.90 = $2042.60
June 17	$120.90	5	5 × $120.90 = $604.50
June 22	$242.85	11	11 × $242.85 = $2671.35
Total		30	$5318.45

Average Daily Balance = $\frac{sum\ of\ daily\ balances}{days\ in\ billing\ cycle} = \frac{5318.45}{30} = \$177.28$

Interest = Monthly Rate × Average Daily Balance

Interest = $\frac{0.18}{12} \times 177.28 = 0.015 \times 177.28 = \$2.66$

13.

Date	Balance Due	Number of Days until Balance Changed	(balance due) × (number of days)
March 1	$35.95	4	4 × 35.95 = 143.80
March 5	$0	7	7 × 0 = 0
March 12	$545.75	20	20 × 545.75 = 10,915.00
Total		31	$11,058.80

Average Daily Balance = $\dfrac{\text{sum of daily balances}}{\text{days in billing cycle}} = \dfrac{11{,}058.80}{31} = \$356.74$

Interest = Monthly Rate × Average Daily Balance

Interest = 0.015 × 356.74 = $5.35

15.

Date	Balance Due	Number of Days until Balance Changed	(balance due) × (number of days)
October 5	$550.77	0	
October 5	$956.75	7	7 × $956.75 = $6697.25
October 12	$606.75	18	18 × $606.75 = $10921.50
October 30	$651.75	6	6 × $651.75 = $3910.50
Total		31	$21,529.25

Average Daily Balance = $\dfrac{\text{sum of daily balances}}{\text{days in billing cycle}} = \dfrac{21{,}529.25}{31} = \$694.49$

Interest = Monthly Rate × Average Daily Balance

Interest = $\dfrac{0.18}{12} \times 694.49 = 0.015 \times 694.49 = \$10.42$

17. Finance charge per hundred = $\dfrac{968.05}{9500} \times 100 = 10.19$

Using the APR table on p. 176, find the closest amount on the 18-month row to $10.19. This amount is $10.19 which gives an interest rate of 12.5% APR

Fundamentals of Algebraic Modeling, 5th Ed.

19. Finance charge per hundred = $\dfrac{32}{1000} \times 100 = 3.20$

   Using the APR table on p. 176, find the closest amount on the 6-month row to $3.20. This amount is $3.23 which gives an interest rate of 11.0% APR.

21. Finance charge per hundred = $\dfrac{1841}{7500} \times 100 = 24.55$

   Using the APR table on p. 176, find the closest amount on the 60-month row to $24.55. This amount is $24.55 which gives an interest rate of 9.0% APR.

23. Total Cost = Payments x Number of Months = $56.82 x 24 = $1363.68

   Finance Charge = Total Cost - Cash Price = $1363.68 - $1250 = $113.68

   Finance charge per hundred = $\dfrac{113.68}{1250} \times 100 = 9.09$

   Using the APR table on p. 176, find the closest amount on the 24-month row to $9.09. This amount is $9.09 which gives an interest rate of 8.5% APR.

25. Use the APR table to find the amount per hundred that she will pay. Go to row 24 and column 8.5% to find $9.09.

   Total Interest Due = $\dfrac{amount\ financed}{100} \times amount\ in\ chart = \dfrac{3560}{100} \times 9.09 = \$323.60$

27. First, subtract the down payment to find the amount financed.
   Down Payment = 10% x 10,876 = 0.10 x 10876 = $1087.60
   Amount financed = Cash Price - Down Payment = $10,876 - $1087.60 = $9788.40

   Use the APR table to find the amount per hundred that she will pay. Go to row 48 and column 12% to find $26.40.

   Total Interest Due = $\dfrac{amount\ financed}{100} \times amount\ in\ chart = \dfrac{9788.40}{100} \times 26.40 = \$2584.14$

   Needed to pay the loan = balance due + interest = $9788.40 + $2584.14 = $12,372.54

   Monthly Payments = $\dfrac{12{,}372.54}{48} = \$257.77/month$

29. First, subtract the down payment to find the amount financed.
   Amount financed = Cash Price - Down Payment = $4578 - $500 = $4078

   Use the APR table to find the amount per hundred that she will pay. Go to row 36 and column 10.5% to find $17.01.

Total Interest Due = $\dfrac{\text{amount financed}}{100} \times \text{amount in chart} = \dfrac{4078}{100} \times 17.01 = \$693.67$

Needed to pay the loan = balance due + interest = 4078 + 693.67 = \$4771.67

Monthly Payments = $\dfrac{4771.67}{36} = \$132.55/\text{month}$

Practice Set 4-4

1. Simple interest is calculated on the full amount borrowed for the full term of the loan. Then interest plus principal are added and divided into equal monthly payments.
3. lower payments, repairs under warranty, (many other possible answers)
5. you may sell the car at any time, you may customize it, (other answers are possible)
7. 46,090 + 5%(46,090) = \$48,394.50
9. 29,460 + 6%(29,460) = \$31,227.60
11. 56,904 + 7%(56,904) = \$60,887.28
13. $I = Prt = (37,884)(0.029)(5) = \$5493.18$
    Pmt = (37,884.00 + 5493.18)/60 = \$722.96 monthly
15. $I = Prt = (15,747)(0.05)(3) = \$2362.05$
    Pmt = (15,747.00 + 2362.05)/36 = \$503.03 monthly
17. $I = Prt = (50,187)(0.019)(6) = \$5721.32$
    Pmt = (50,187 + 5721.32)/72 = \$776.51 monthly
19. cost = (317.50)(36) = \$11,430.00
    Interest = 11,430 − 9000 = \$2430.00
21. cost = 3290 + (339.97)(48) = \$19,608.56
    Interest = 19,608.56 − 16,450 = \$3158.56
23. $I = 36(317.50) - 9000 = \$2430.00$

    $r = \dfrac{I}{Pt} = \dfrac{2430}{9000(3)} = 0.09 = 9\%$

25. $I = 48(339.97) - (16450 - 3290) = \$3158.56$

    $r = \dfrac{I}{Pt} = \dfrac{3158.56}{13160(4)} = 0.06 = 6\%$

27. $I = 500 + 127.20(36) - 4500 = 579.20$

$\dfrac{579.20}{4000} \times 100 = 14.48$; find closest to 14.48 on row 36 of the APR table

APR = 9.0%

29. APR table row 36, 10.5% = $17.01 per $100 financed

Amt. Financed = 8578 − 500 = $8078

$I = 17.01\left(\dfrac{8078}{100}\right) = \$1374.07$

Pmt = (8078 + 1374.07)/36 = $262.56

31. $A = 426.71 + 48(426.71) + 895 = \$21{,}803.79$

33. $25{,}000 - 10\%(25{,}000) = \$22{,}500$

Pmt = $\dfrac{P+I}{48} = \dfrac{P+Prt}{48} = \dfrac{22{,}500 + 22{,}500 \times .029 \times 4}{48} = \$523.13$

Total cost = 2500 + 48(523.13) = $27,610

35. Extra miles = 65,200 − 4(15,000) = 5200 miles

5200(0.18) = $936.00 extra

Practice Set 4-5

1. interest rate, fees, points, type of loan, etc.

3. subtract your monthly bills from your gross income and multiply the balance by 36%

5. use the maximum amount of loan formula and the amortization table (page 189)

7. down pmt = 10%(153,250) = $15,325; amt. fin = 153,250 − 15,325 = $137,925

9. down pmt = 25%(179,900) = $44,975; amt. fin = 179,900 − 44,975 = $134,925

11. down pmt = 30%(315,500) = $94,650; amt. fin = 315,500 − 94,650 = $220,850

13. (gross income − monthly bills)(36%) = (985.65 − 135.50)(36%) = $306.05

15. (2200 − 550)(36%) = $594.00

17. (3700 − 650)(36%) = $1098.00

19. $P = 185{,}000\left[\dfrac{\dfrac{0.0725}{12}\left(1+\dfrac{0.0725}{12}\right)^{12 \times 25}}{\left(1+\dfrac{0.0725}{12}\right)^{12 \times 25}-1}\right] = \$1337.20$

Fundamentals of Algebraic Modeling, 5th Ed.

21. $P = 110{,}500 \left[\dfrac{\dfrac{0.0625}{12}\left(1+\dfrac{0.0625}{12}\right)^{12 \times 30}}{\left(1+\dfrac{0.0625}{12}\right)^{12 \times 30}-1} \right] = \$680.37$

23. $P = 113{,}650 \left[\dfrac{\dfrac{0.0585}{12}\left(1+\dfrac{0.0585}{12}\right)^{12 \times 20}}{\left(1+\dfrac{0.0585}{12}\right)^{12 \times 20}-1} \right] = \$804.42$

25. $P = \dfrac{85{,}000}{1000} \times 6.75 = \$573.75$

27. $P = \dfrac{112{,}000}{1000} \times 8.99 = \$1006.88$

29. $P = \dfrac{265{,}000}{1000} \times 6.32 = \$1674.80$

31. max loan $= \dfrac{750}{6.32} \times 1000 = \$118{,}670.89$

33. max loan $= \dfrac{1850}{7.07} \times 1000 = \$261{,}669.02$

35. max loan $= \dfrac{635}{7.69} \times 1000 = \$82{,}574.77$

37. down pmt = 25%(250,000) = \$62,500; amt. fin. = 250,000 − 62,500 = \$187,500

$P = 187{,}500 \left[\dfrac{\dfrac{0.0575}{12}\left(1+\dfrac{0.0575}{12}\right)^{12 \times 30}}{\left(1+\dfrac{0.0575}{12}\right)^{12 \times 30}-1} \right] = \$1094.20$

39. a) down pmt = 15%(118,500) = \$17,775; balance = \$100,725

$P = 100{,}725 \left[\dfrac{\dfrac{0.0625}{12}\left(1+\dfrac{0.0625}{12}\right)^{12 \times 30}}{\left(1+\dfrac{0.0625}{12}\right)^{12 \times 30}-1} \right] = \$620.19$

b) A = 360(620.19) + 17,775 = \$241,043.40

41. go to www.aliaterra.com

Fundamentals of Algebraic Modeling, 5th Ed.

Practice Set 4-6

1. An economical way of helping an individual deal with potential severe financial losses by pooling money with a large group of people.

3. The total amount of insurance specified by the insurance policy purchased is the face value.

5. The payment made to the policyholder in the event of an insured loss is called an indemnity.

7. $20,000 for bodily injury per person injured, $40,000 bodily injury per accident, and $15,000 property damage per accident (20/40/15)

9. Collision insurance pays for repairs to your own car if you are responsible for an accident.

11. $101 + 99 = 200; 200(1.50) = \$300$

13. $91 + 95 = 186; 186(1.10) = \$204.60$

15. $136 + 107 = 243; 243(2.60) = \$631.80$

17. $57 + 94 = 151; 151(1.40) = \$211.40$

19. $52 + 73 = 125; 125(1.40) = \$175.00$

21. $64 + 125 = 189; 189(1.40) = \$264.60$

23. $(86 + 83 + 55 + 82)(1.90) = \$581.40$

25. a) 8000

    b) $0

Practice Set 4-7

1. a) $57.15(1500) = \$85,725$

   b) $0.02(85,725) = \$1714.50$

3. a) $27.33(50) = \$1366.50$

   b) $0.015(1366.50) + 0.10(50) = \$25.50$

5. a) $37.80(1100) = \$41,580$

   b) $0.021(41,580) = \$873.18$

7. a) $24.74(40) = \$989.60$

   b) $0.025(989.60) + 0.125(40) = \$29.74$

9. a) $500(0.40) = \$200$

   b) $178,606(100) = 17,860,600$

   c) $29.90

   d) $29.9 + 0.37 = \$30.27$

   e) $0.37/30.27 = 1.22\%$ decrease

11. a) 50(0.32) = $16

 b) 68,540(100) = 6,854,000

 c) $22.72

 d) 22.72 − 0.04 = $22.68

 e) 0.04/22.68 = 0.18%

13. 2155 ÷ 8.62 = 250 shares

15. 10,150 ÷ 29 = 350 shares

17. Bonds are issued by companies and governments to raise money for projects. Bonds are generally safer investments than stocks.

19. Invest more of your funds in stocks and mutual funds since, over long periods of time, these types of investments have a history of higher growth rates than bonds.

Practice Set 4-8

1. P = ($15670)(9.5%) = ($15670)(0.095) = $1488.65

3. P = $1245 + (3.5%)($65000) = $1245 + $2275 = $3520

5. P = 2200 + (2%)(12350 - 7500) = 2200 + (0.02)(4850) = 2200 + 97 = $2297

7. P = (2.5%)(125,000) = (0.025)(125,000) = $3125

9. Tip = 15%(65) = 0.15(65) = $9.75

11. 0.06(150,000) = 9000; 0.60(9000) = $5400 commission

13. 425/5000 = 0.085 = 8.5%

15. a) tax = 1275 + 7%(48745 - 21250) = $3199.65

 b) tax = 765 + 7%(48745 - 12750) = $3284.65

17. Fixed rate tax: 4.5%(31,000) = 0.045(31,000) = $1395

 Tax Table: 765 + 0.07(22,450-12,750) = 765 + 0.07(9700) = $1444

 He owes more if he uses the tax table.

19. Income = 0.045(12)(50,000) = $27,000

 Tax = 765 + 0.07(16,750 − 12,750) = $1045

21. a) C = 800 + 9.5%(S) = 800 + 0.095S

 b) The above equation is written in slope-intercept form. slope = m = 9.5% = 0.095 dollars per sale

 c) Slope units would be dollars of commission earned per dollars of sales made

 d) Yes because there is only one commission for each amount of sales made.

e) "Commission" would go on the y-axis since it is dependent upon the amount of sales.

f) $2525 = 800 + 9.5\%(S)$

$2525 - 800 = 0.095S$

$1725 = 0.095S$

$S = \$18157.89$ (rounded to the nearest cent)

23. Housing   Rent $590

                Utilities   25

                Phone   45

Loans   Car   125

                Food   320

     Entertainment   150

                  TV   45

             Gasoline   150

Total Budget Amt. = $1450

    Savings desired = $0.07(1855) = \$129.85$

The total budgeted plus the desired savings amount is less than the total take-home pay, so saving 7% is possible.

25. $0.15(1978) = \$296.70$

27. a) 36%(income – expenses) = estimated maximum loan payment

    $0.36(4250 - 215 - 45 - 55 - 45) = \$1400.40$

b) taxes = $0.15(4250) = 637.50$

  utilities = $0.0425(4250) = 180.63$

  food = $0.135(4250) = 573.75$

  savings = $0.055(4250) = 233.75$

  entertainment = $0.04(4250) = 170.00$

  gasoline/car = $0.05(4250) = 212.50$

  insurance = $0.02(4250) = 85.00$

  Total Budget Amount = $3853.13 (including house payment and monthly expenses)

  This leaves $396.47 per month out of the gross pay.

29. Answers will vary.

Chapter 4 Review Problems

1. County tax = 0.025(12,548) = $313.70

   State tax = 0.045(12,548) = $564.66

2. x + 8.5%(x) = 4285.75

   $\qquad$ 1.085x = 4285.75

   $\qquad\quad$ x = $3950

3. 0.05(12.80) + 0.06(26.50) = $2.23

4. tip = 15 − 12.50 = $2.50; 2.50/12.50 = 0.2 = 20%

5. straight line depreciation amt. = (25,000 − 1500)/10 = $2,350

6. Profit = gross sales - costs = 75(275) - [1250 + 75(125)] = $10,000.00

7. (200%)(475) = 2(475) = $950 markup; selling price = 950 + 475 = $1,425

8. Commission only: 25%(5000) = $1250.00

   Salary plus commission: 1000 + 5%(5000) = 1000 + 250 = $1250.00

   So for July, Julian would make the same either way.

9. $I = Prt$; $r = \dfrac{I}{Pt} = \dfrac{162.50}{(5000)(0.5)} = 0.065 = 6.5\%$ interest rate

10. part 1: $I = Prt = 15000(13\%)(2) = \$3900$ and $M = P + I = 15000 + 3900 = \$18,900$

    part 2: payments = 18900/(2·12) = 18900/24 = $787.50 per month

11. $I = Prt = 1000(0.06)(6) = \$360$

12. $r = 5.8\%$ and $t = 2$

    $$M = 8000\left(1 + \frac{0.058}{4}\right)^{4(2)} = 8000(1.122060854) = 8976.4868\ldots = \$8976.49$$

13. Find the compound amount, M, and subtract the principal, P, to get the interest that was added.

    $$I = M - P = 2500\left(1 + \frac{0.05}{12}\right)^{12(5)} - 2500 = 2500(1.283358\ldots) - 2500 = 3208.40 - 2500 = \$708.40$$

14. $A = 10,000e^{(0.08)(20)} = 49,530.32424 = \$49,530.32$

15. $3150.75\left(\dfrac{0216}{12}\right) = \$56.71$

16.

Date	Balance Due	Number of Days until Balance Changed	(balance due) × (number of days)
June 1	$1415.90	14	14 × $1415.90 = 19,822.60
June 15	$1290.90	8	8 × $1290.90 = 10,327.20
June 23	$1312.85	8	8 × 1312.85 = 10,502.80
Total		30	$40,652.60

Average Daily Balance = $\dfrac{\text{sum of daily balances}}{\text{days in billing cycle}} = \dfrac{40{,}652.60}{30} = \$1355.09$

Interest = Monthly Rate x Average Daily Balance

Interest = $\dfrac{0.18}{12} \times 1355.09 = 0.015 \times 1355.09 = \$20.33$

17. APR table row 36, 8% = $12.81 per $100 financed

   Amt. Financed = 2350 − 500 = $1850

   I = $12.81\left(\dfrac{1850}{100}\right) = \$236.99$

   Pmt = (1850 + 236.99)/36 = $57.98

18. 26,998 + 0.05(26,998) = $28,347.90

19. $I = Prt = 27{,}885(0.009)(5) = \$1254.83$

   Pmt = (27,885 + 1254.83)/60 = $485.67

20. cost = 36(196.80) = $7084.80

   I = 7084.80 − 6500 = $584.80

   r = $\dfrac{584.80}{6500(3)} = 0.0299\ldots = 3\%$

21. 395.59 + 849 + 48(395.59) = $20,232.91

22. 62,148 − 4(15,000) = 2148 miles

   Extra cost = (2148 miles)($0.15/mile) = $322.20

23. Cost = 25,000 + 25,000(0.036)(5) = $29,500

   Pmt = 29,500/60 = $491.67 monthly

24. 36%(income − expenses) = estimated maximum loan payment

   0.36(4988 − 500.30) = $1615.57

25. Balance to finance = 145000−(20%)(145000)=$116,000

$$P = 116000\left(\dfrac{\dfrac{0.0625}{12}\left(1+\dfrac{0.0625}{12}\right)^{12*30}}{\left(1+\dfrac{0.0625}{12}\right)^{12*30}-1}\right) = 116000\left(\dfrac{0.0337977416}{5.48916638}\right) = \$714.24$$

26. $P = \dfrac{148{,}500}{1000} \times 5.68 = \$843.48$

27. (64 + 104)(1.40) = $235.20

28. (129 + 108)(1.40) = $331.80

29. $20,000

30. 150(36.54) + 10.99 = $5491.99

31. a) 450(27.28) = $12,276

    b) 0.025(12,276) = $306.90 commission

32. 87250/34.90 = 2500 shares

33. x(12500) = 812.50

    x = 812.50/12500 = 0.065 = 6.5% commission

34. Salary = 400 + 6.5%(6000−1200) = 400 + 0.065(4800) = $712.00

35. T = $1275 + 0.07(x − 21250)

36. T = 1275 + 7%(22500−21250) = 1275 + 87.5 = $1362.50

37. 0.15(2386) = $357.90

Chapter 4 Test

1. Tax = 0.035(24,500) = $857.50

2. 12.40 + 1.25(12.40) = $27.90

3. Depreciation amount = (18,000 − 2500)/10 = $1550.00 per year

4. P = 14,500(2.20) − [3575 + 14500(0.82)] = $16,435.00

5. P = 200 + 0.12(1275) = $353.00

6. T = 18%(55,250) = 0.18(55,250) = $9945.00

7. $M = 2500\left(1+\dfrac{0.045}{4}\right)^{4(15)} = 2500(1.956645179) = 4891.61$

8. 250/500 = 0.05 = 5%

9. $M = 7500e^{(0.055)(6)} = 10432.26096 = \$10{,}432.26$

10. 5250.05(0.15/12) = $65.63

Fundamentals of Algebraic Modeling, 5th Ed.

11.

Date	Balance Due	Number of Days until Balance Changed	(balance due) ×(number of days)
July 1	$115.90	14	14 × 115.90 = 1622.60
July 15	$90.90	5	5 × 90.90 = 454.50
July 20	$316.45	12	12 × 316.45 = 3797.40
Total		31	$5874.50

$$\text{Average Daily Balance} = \frac{\text{sum of daily balances}}{\text{days in billing cycle}} = \frac{5874.50}{31} = \$189.50$$

Interest = Monthly Rate × Average Daily Balance

$$\text{Interest} = \frac{0.18}{12} \times 189.50 = 0.015 \times 189.50 = \$2..84$$

12. Amount to finance = 1385 − 100 = $1285

$$P = 9.64 \left(\frac{1285}{100} \right) = \$123.87$$

   Pmt = (1285 + 123.87)/24 = $58.71

13. $I = Prt = 57{,}885(0.009)(6) = \$3125.79$

   Pmt. = (57,885 + 3125.79)/72 = $847.38

14. Cost = 571 + 54(571) + 795 = $32,200

15. Amount to finance = 36,200 − 600 = $35,600

   I = Prt = 35,600(0.019)(6) = $4058.40

   Pmt. = (35,600 + 4058.40)/72 = $550.82

16. Amount to finance = 375,000 − 0.25(375,000) = $281,250

   Table 4.4, row 5.5%, 30 years, = 5.68

$$P = \frac{281{,}250}{1000} \times 5.68 = \$1597.50$$

17. (103 + 101)(1.50) = $306

18. $0

19. 175(49.5) + 250(29.88) + 2(9.99) = $16,152.48

20. a) 50(44.50) = $2225

   b) Commission = 0.015(2225) + 0.15(50) = $40.88

Chapter Five

Practice Set 5-1

1. A square has 4 equal sides. P = 4L = 54 in., so L = 13.5 in per side. The area of a square is given by the formula: $A = L^2 = (13.5 \text{ in})^2 = 182.25 \text{ in}^2$.

3. Use the formula for the area of a rectangle to calculate the size of the area to be planted. $A = lw = (8.75 \text{ ft})(4.5 \text{ ft}) = 39.375 \text{ ft}^2$. Since she needs four plants for every square foot, you should multiply this answer by 4. This gives $39.375 \times 4 = 157.5$ plants. In a practical sense, using this answer means 158 plants. If you round the area off to 40 ft² and multiply by 4, you would get 160 plants.

5. The perimeter minus the door openings will equal the number of feet of molding needed: 3 + 3 + 2 (the offset) + 17 + 12 + 4 + 19 + 14 = 74 feet of molding. To find the area of the room, divide it into two rectangles. We chose to divide it at the offset to the right of the door. This gives a rectangle that is 14 ft by 9 feet (including the door width for carpet) and one that is 17 ft by 12 ft. So, the total area is:

    $A = (14)(9) + (17)(12) = 330 \text{ ft}^2$ Then, 330 ft² divided by 9 square feet per square yard gives 36.666 yd² = 37 yd² of carpet.

7. Hero's formula: s = (18 + 11 + 25)/2 = 27

    $A = \sqrt{27(27-18)(27-11)(27-25)} = \sqrt{7776} = 88.18 \text{ ft}^2$

9. To surround a circular walkway, we need to find the circumference: $C = \pi d$. For this problem, since the radius is 25 feet, the diameter (d) is 50 feet. Therefore, $C = \pi d = \pi \cdot 50 = 157$ feet. Since plants should be planted every six inches, that means you need 2 plants/foot. Therefore you will need (2 plants/ft)(157 ft) = 314 plants.

11. The label is a rectangle with one side being the height of the can and the other the circumference of the circular end of the can.

    $A = lw = (4.5 \text{ in})(7.9 \text{ in}) = 35.55 \text{ in}^2$ of paper

13. The paper is rectangular so it will cover an area of $A = lw = (20 \text{ in})(15 \text{ in}) = 300 \text{ in}^2$. The top, sides and bottom of the box must be covered so we will calculate the area of each side of the box and total the area.

    $$\text{Top} = lw = (12 \text{ in})(8 \text{ in}) = 96 \text{ in}^2$$
    $$\text{Side} = lw = (12 \text{ in})(3 \text{ in}) = 36 \text{ in}^2$$

$$\text{End} = lw = (8 \text{ in})(3 \text{ in}) = 24 \text{ in}^2$$

For the whole box, there are six surfaces, (top, bottom, 2 ends, 2 sides) so the total area is $2(96 \text{ in}^2) + 2(36 \text{ in}^2) + 2(24 \text{ in}^2) = 312 \text{ in}^2$. Therefore, this piece of paper is not big enough to cover the box.

15. Divide the areas by the cost for each pizza to find the less costly pizza; $A = \pi r^2$ for each pizza.

    8 in diameter pizza: $A = \pi(4 \text{ in})^2 = 50.265 \text{ in}^2$ divided by 699 cents = 0.0719 in²/cent

    16 in diameter pizza: $A = \pi(8 \text{ in})^2 = 201.062 \text{ in}^2$ divided by 1595 cents = 0.126 in²/cent

    So, you get more square inches of pizza to eat per cent with the 16-inch pizza.

17. Find the area of the square. Then find the area of the circle and subtract the areas.

    $$\text{Area of the square} = s^2 = (24 \text{ m})^2 = 576 \text{ m}^2$$

    $$\text{Area of the circle} = \pi r^2 = \pi \cdot 12^2 = \pi \cdot 144 = 452.39 \text{ m}^2$$

    $$\text{Area of the shaded part} = 576 \text{ m}^2 - 452.39 \text{ m}^2 = 123.61 \text{ m}^2$$

19. a) cost divided by the area = $\dfrac{\$8625}{(50 ft)(230 ft)} = \dfrac{\$8625}{11500 ft^2} = \$0.75/ft^2$

    b) $\$8625/50$ ft = $\$172.50$ per foot of frontage

    c) $(43560 \text{ ft}^2/\text{acre})(\$0.75/\text{ft}^2) = \$32,670/\text{acre}$

    or, the hard way: $43560/11500 = 3.787826087$ lots of the size purchased to equal one acre; $\$8625(3.787826087) = \$32,670$ for an acre at the same cost per square foot.

21. $A = l^2$ and $P = 4l$

    a) $A = (2l)^2 = 4l^2$ or 4 × original area and $P = 4(2l) = 8l$ or 2 × original perimeter

    b) $A = (0.5l)^2 = 0.25l^2$ or ¼(original area) and $P = 4(0.5l) = 2l$ or ½ of original perimeter.

    c) $A = (3l)^2 = 9l^2$ or 9 × original area and $P = 4(3l) = 12l$ or 3 × original perimeter

    d) Doubling the length of the diagonal of a square will double the side lengths of the square; thus, the results will be the same as in a) above.

23. The line formed by the diameters of four circles in a row with their sides touching each other and the sides of the square will equal the length of the rectangle and the line formed by two diameters of circles in a column will equal the width of the rectangle.

W = 2d

L = 4d

Find the total area of 8 equal circles based on a diameter of d for each (r = 0.5d) then divide that by the area of a rectangle with L = 4d and W = 2d.

$A_{circle} = \pi r^2 = \pi(0.5d)^2 = 0.25\pi d^2 = 0.785...d^2$

so 8 circles have an area of about $6.283d^2$

$A_{rectangle} = lw = (4d)(2d) = 8d^2$

$$\frac{A_{circle}}{A_{rectangle}} = \frac{6.283d^2}{8d^2} = 0.785 = 78.5\%$$

25. a) We started by multiplying the formula given by 2 to clear the fraction.

$2w + 4h + \pi w = 40$ ft

$4h = 40$ ft $- 2w - \pi w$

$$h = \frac{40 \text{ ft}}{4} - \frac{2w}{4} - \frac{\pi w}{4}$$

h = 10 ft - 0.5w - 0.785w = 10 ft - 1.285w or 10 ft - 1.3w

b) Base the graph on the intercepts, h = 0 and w = 0

0 = 10 - 1.3w

w = 7.7, which gives one intercept, (7.7 , 0)

h = 10 - 1.3(0)

h = 10, which gives the other intercept, (0 , 10)

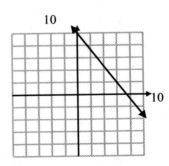

c) We chose for w to be simple lengths or 0, 1, 2, and 3 feet, then solved for h.

for w = 0, h = 10 - 1.3w = 10 - 1.3(0) = 10

for w = 1, h = 10 - 1.3(1) = 8.7

for w = 2, h = 10 - 1.3(2) = 7.4

for w = 3, h = 10 - 1.3(3) = 6.1

d) Theoretically one could choose any positive real number to represent the width of the window and this would lead to an infinite number of possibilities.

27. $d = 8$ft, $r = 4$ft; $A = \pi r^2 = \pi 4^2 = 50.265\ldots = 50.3$ ft^2

29. $A_{circle} - A_{pool} = \pi 6^2 - \pi 4^2 = 62.8$ ft^2

(4 pavers per square foot)(62.8ft^2) = 251.2 = 252 pavers needed

Cost = (252)($0.59) = $148.68

Practice Set 5-2

1. The triangle must be a right triangle and you must know the length of two sides.

Note: for problems 3 – 8 the Pythagorean Theorem is used with c being the hypotenuse.

3. $c^2 = 9^2 + 12^2$
   $c^2 = 81 + 144$
   $c^2 = 225$
   $c = \sqrt{225} = 15.0$

5. $12^2 = a^2 + 6^2$
   $12^2 - 6^2 = a^2$
   $144 - 36 = a^2$
   $108 = a^2$
   $\sqrt{108} = 10.4 = a$

7. $26^2 = 10^2 + b^2$
   $b^2 = 676 - 100$
   $b^2 = 576$
   $b = \sqrt{576} = 24$

9. The ladder length will be the hypotenuse:
   $25^2 = h^2 + 7^2$
   $h^2 = 625 - 49 = 576$
   $h = \sqrt{576} = 24$ ft

11.

Use the Pythagorean Theorem, letting h = a.
$$h^2 + 9^2 = 13^2$$
$$h^2 = 13^2 - 9^2$$
$$h^2 = 169 - 81 = 88$$
$$h = \sqrt{88} = 9.4 \ m$$

13.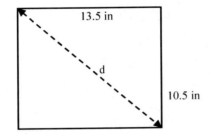

Use the Pythagorean Theorem, letting d = c.
$$13.5^2 + 10.5^2 = c^2$$
$$182.25 + 110.25 = c^2$$
$$292.5 = c^2$$
$$\sqrt{292.5} = 17.1 = c$$

This is a 17 in screen.

15.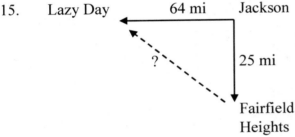

$$a^2 + b^2 = c^2$$
$$64^2 + 25^2 = c^2$$
$$4096 + 625 = c^2$$
$$4721 = c^2$$
$$\sqrt{4721} = 68.7 mi = c$$

17.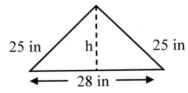

The height (h) divides the triangle into two right triangles with bases of 14 in. Use Pythagorean Theorem to find h.

$$a^2 + b^2 = c^2$$
$$h^2 + 14^2 = 25^2$$
$$h^2 + 196 = 625$$
$$h^2 = 625 - 196 = 429$$
$$h = \sqrt{429} = 20.7 in.$$

Fundamentals of Algebraic Modeling, 5th Ed.

19. 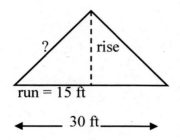 The slope is given as 11:20. Remember that by definition this is rise over run. Use the slope to find the rise of this roof and then use the Pythagorean Theorem to find the missing side.

$$\frac{11}{20} = \frac{rise}{15}$$
$$(15)(11) = 20x$$
$$165 = 20x$$
$$8.25 = x \text{ (rise)}$$

$$a^2 + b^2 = c^2$$
$$8.25^2 + 15^2 = c^2$$
$$293.0625 = c^2$$
$$\sqrt{293.0625} = 17.12 \; ft = c$$

21. Use the Pythagorean theorem to find the base length:
$$a^2 + b^2 = c^2$$
$$10^2 + b^2 = 26^2$$
$$b^2 = 676 - 100 = 576$$
$$b = \sqrt{576} = 24 \text{ cm.}$$

Perimeter: $P = 10 + 26 + 24 = 60$ cm

Area: $A = \frac{1}{2}(24)(10) = 120 \text{ cm}^2$

23. The length of side AC is 8 units, and the length of side BC is 4 units. Use the Pythagorean Theorem to find the length of AB.
$$a^2 + b^2 = c^2$$
$$8^2 + 4^2 = c^2$$
$$64 + 16 = c^2$$
$$80 = c^2$$
$$\sqrt{80} = 8.94 = c$$

25. The doorway is a rectangle. In order for the mirror to fit through the doorway, it would be turned diagonally to pass through the doorway. If the diameter of the mirror is longer than the diagonal of the doorway, it will not fit.

$$a^2 + b^2 = c^2$$
$$200^2 + 75^2 = c^2$$
$$40{,}000 + 5625 = c^2$$
$$45625 = c^2$$
$$\sqrt{45625} = 213.6 = c$$

Since the mirror has a 220 cm diameter and the doorway is only 213.6 cm on the diagonal, the mirror will not go through the door.

27. The diagonal will be the hypotenuse of an isosceles right triangle with sides s.

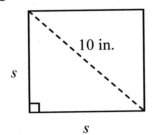

$$a^2 + b^2 = c^2$$
$$s^2 + s^2 = 10^2$$
$$2s^2 = 100$$
$$s^2 = 50$$
$$s = \sqrt{50} = 7.07 \text{ in}$$

Practice Set 5-3

1. To make drawings in 2-dimensions that are true representations of what we see in 3-dimensions with our eyes.

3. The line on which the principal vanishing point in a perspective drawing is located.

5. The term *symmetry* implies that balance or a regular pattern exists.

7. (various drawings are possible here)

9. (various drawings are possible here)

11. (various answers are possible here)

13. (various answers are possible here)

15. (various answers are possible here but they should be similar to fig. 5-17)

17. horizontal symmetry: B, C, D, E, H, I, K, O, X.

    vertical symmetry: A, H, I, M, O, T, U, V, W, X, Y.

    F, G, J, L, N, P, Q, R, S, and Z do not have horizontal or vertical symmetry

Fundamentals of Algebraic Modeling, 5th Ed.

19. H, I, N, O, S, X, Z.

21. one-third of a full rotation = 120°

23. one-fifth of a full rotation = 72°

25. An octagon will have 8 equal sides;

formula: $\dfrac{360°}{\text{number of sides}} = \dfrac{360°}{8} = 45°$

27.

Sketch from front

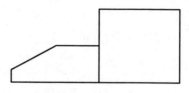
Sketch from right hand side

Practice Set 5-4

1. $\left(\dfrac{16\,ft}{1\,in}\right)(21\,in) = 336\,ft$

3. $\left(\dfrac{3}{4}\right)(175\,cm) = 131.25\,cm$

5. $\left(\dfrac{800\,cm}{1\,cm}\right)(42.2\,cm) = 33{,}760\,cm = 337.6\,m$ long

   $\left(\dfrac{800\,cm}{1\,cm}\right)(8.2\,cm) = 6560\,cm = 65.6\,m$ wide

7. $(8cm)\left(\dfrac{5000}{1}\right) = 40{,}000\,cm = 400\,m$

9. approximate length of photo = 2.75 in

   $\dfrac{573\,ft}{2.75\,in} = 208.363\,ft$ or 1 in = 208 ft

(answers may vary slightly depending upon accuracy of measurement)

11. $(600 \text{ } mi)\left(\dfrac{0.5 \text{ } in}{50 \text{ } mi}\right) = 6 \text{ } in$

13. (student responses will vary)

15. from the text, Fibonacci #15 = 610 and #16 = 987, so

   #17 = 610 + 987 = 1597

   #18 = 987 + 1597 = 2584

   #19 = 1597 + 2584 = 4181

   #20 = 2584 + 4181 = 6765

17. The decimal portions of the two numbers are the same. ($\phi^2 = \phi + 1$)

19. (23 ft)(1.62) = 37.26 ft = 37 ft 3.1 in

21. (45.5 cm)(1.62) = 73.7 cm; $\dfrac{45.5 \text{ } cm}{1.62} = 28.1 \text{ } cm$

23. $\dfrac{w}{h} = \dfrac{l}{w} \approx 1.62$

   $\dfrac{w}{5 \text{ } in} = 1.62$; w = 8.1 in

   $\dfrac{l}{8.1 \text{ } in} = 1.62$; l = 13.1 in

Practice Set 5-5

1. Pitch refers to how our ears interpret different frequencies of sound. The higher the frequency of the sound wave, the higher the pitch we hear.

3. A recording in which the actual wave form is copied as closely as physically (technologically) possible.

5. Frequencies are measured numerically and recorded as numbers in short strings.

7. One octave higher = (261.626 Hz)(2) = 523.252 Hz

   One octave lower = (261.626 Hz)/2 = 130.813 Hz

9. (392 Hz)(1.05946) = 415.3 Hz

11. 4/4 time = 4 quarter notes and each 2 quarter notes = 1 half note, so 2 half notes per measure.

   Or 4/4 time = $\dfrac{4}{4} = \dfrac{2}{4} + \dfrac{2}{4} = \dfrac{1}{2} + \dfrac{1}{2} = 2$ half notes

13. 3/2 time = 3 halves or 3 half notes per measure; $\frac{3}{2} = 3\left(\frac{1}{2}\right) = 3$ half notes.

15. $\frac{3}{4} = \frac{1}{4} + \frac{1}{4} + x$; x = 1 quarter note missing.

17. $\frac{3}{8} = \frac{1}{8} + \frac{1}{8} + x = \frac{2}{8} + x$; x = 1 eighth note missing

19. $\frac{2}{4} = \frac{1}{2} + x$; x = 0 notes missing

21. y = 5 sin(220t); (similar to fig. 5.25 or 5.26)

23. y = 5 sin(196t); (similar to fig. 5.25 or 5.26)

25. (group project results will vary)

Chapter 5 Review Problems

1. Use Hero's formula: $s = \frac{6+8+12}{2} = \frac{26}{2} = 13$;

   $A = \sqrt{13(13-6)(13-8)(13-12)} = \sqrt{13(7)(5)(1)} = \sqrt{455} = 21.3 in^2$

2. P = 10 + 25 + 6 + 16 + (10 - 6) + (25 - 16) = 70 m
   A = (10)(25 - 16) + (16)(6) = 186 m^2

3. amount of molding = L + H + H + half the circumference of a circle for the curved top

   amt. of molding = 5 + 4.5 + 4.5 + ½(π)(5) = 21.85398 ft

   cost = ($0.55)(21.85398) = 12.01968 = $12.02

4. Find the area of the circular sidewalk and tree then subtract the area of the tree in the center: A = π(9.5)^2 - π(1.5)^2 = 276.5 ft^2

5. (a) Here we break the figure into the area of a rectangle plus the area of a triangle by Hero's formula minus the area of each window. For Hero's formula:

   $s = \frac{25+25+42}{2} = 46$

   $A = 42(30) + \sqrt{46(46-25)(46-25)(46-42)} - 3(3.5) - 3(3.5) - (9)(3.5) = 1492.4\, ft^2$

   (b) 1 yd^2 = 9 ft^2; $\frac{1492.4}{9} = 165.82 = 166$ yd^2

   166($22.25) = $3693.50

6. The length of the ladder, L, will be the hypotenuse of the right triangle.

    $L^2 = 15^2 + 8^2 = 289$

    $L = \sqrt{289}$

    $L = 17$ feet

7. The first base to third base distance, d, across the middle of the infield will be the hypotenuse of a right triangle.

    $d^2 = 60^2 + 60^2 = 7200$

    $d = \sqrt{7200}$

    $d = 84.8528 = 84.9$ feet

8. If these side lengths do indeed form a right triangle, the longest side must be the hypotenuse and the lengths of the other sides must correctly fulfill the Pythagorean theorem. So does $12.5^2 = 10^2 + 7.5^2$? $12.5^2 = 156.25$ and $10^2 + 7.5^2$ also equals 156.25. Thus, these side lengths do form a right triangle.

9. $s^2 = 12^2 + 18^2 = 468$

    $s = \sqrt{468}$

    $s = 21.633\ldots = 21.6$ in.

10. (many different examples are possible)

11. (many different examples are possible)

12. (many different examples are possible)

13. a) vertical reflection symmetry

    b) horizontal reflection symmetry

    c) vertical reflection symmetry

    d) neither type of symmetry exists here

14. scaling factor = $\dfrac{50 \, cm}{75 \, cm} = \dfrac{2}{3}$; a scaling factor of 2 to 3

15. actual length = $(4.125 \, in)\left(\dfrac{32 \, in}{1 \, in}\right) = 132 \, in = 11 \, ft$

16. the shorter sides = $\dfrac{32.4 \, cm}{1.62} = 20 \, cm$

17. $(35\ mi)\left(\dfrac{1\ in}{20\ mi}\right) = 1.75\ in$

18. Frequency is a measure of the number of vibrations (waves) per second. Pitch is how high or low a frequency sounds to the human ear.

19. One octave lower = $\dfrac{110\ Hz}{2}$ = 55 Hz

    Two octaves lower = $\dfrac{55\ Hz}{2}$ = 27.5 Hz

20. $\dfrac{3}{2} = \dfrac{6}{4} = 6$ quarter notes per measure

Chapter 5 Test

1. By Hero's formula, $s = \dfrac{2.6 + 3.4 + 4.1}{2} = 5.05$

   $A = \sqrt{5.05(5.05 - 2.6)(5.05 - 3.4)(5.05 - 4.1)} = 4.4038 = 4.4\ ft^2$

2. (a) P = 21.5 + (18.5 - 4.25) + (21.5 - 15) + 4.25 + 15 + 18.5 = 80 ft
      A = 18.5(15) + (18.5 - 4.25)(21.5 - 15) = 370.125 ft^2

   (b) 370.125/9 = 41.125 yd^2 (or 42 yd^2 if you buy whole square yards)
      Cost = 41.125(24.95) = 1026.06875 = $1026.07 (or 42(24.95) = $1047.90 for whole square yards)

3. The ladder's length will be the hypotenuse of the right triangle.
   $26^2 = 4.5^2 + b^2$
   $b^2 = 655.75$
   $b = 25.6076 = 25.6$ ft

4. The diagonal of the rectangle will be the hypotenuse of a right triangle:
   $6.5^2 = 5.8^2 + b^2$
   $b^2 = 8.61$
   $b = 2.9342 = 2.9$ cm

5. (a) In one revolution, it will roll a distance equal to the circumference of the circle.
      $C = \pi d = \pi(21) = 65.973\ldots = 65.97$ in or 65.97/12 = 5.498 ft

   (b) The number of revolutions = 5280/5.498 = 961 rev (960.349)

6. The primary vanishing point is located on the horizon line in the drawing.

7. Only letter "O".

8. Letters O and X.

9. $6250 \, m \left(\dfrac{1 \, cm}{2500 \, m} \right) = 2.5 \, cm$

10. $(75 \, ft) \left(\dfrac{1 \, ft}{87 \, ft} \right) = 0.862... \, ft = (0.862 \, ft) \left(\dfrac{12 \, in}{1 \, ft} \right) = 10.3 \, in$

11. $\dfrac{64 \, in}{48 \, in} = 1.333 \neq 1.62$ so, this ratio is not equivalent to the Golden Ratio.

12. scaling up in size: (48 in)(1.62) = 77.8 in

    Scaling down in size: (48 in)/1.62 = 29.6 in

13. Notebook paper ratio = $\dfrac{11 \, in}{8.5 \, in} = 1.29$

    Legal pad ratio = $\dfrac{14 \, in}{8.5 \, in} = 1.65$; The legal pad is closer to the Golden Ratio of 1.62.

14. Analog recordings are essentially copies of the actual physical wave shape whereas digital recordings are numerical in nature.

15. 2(440 Hz) = 880 Hz (one octave)

    2(880 Hz) = 1760 Hz (two octaves)

    2(1760 Hz) = 3520 Hz = three octaves above 440 Hz

16. Sound and light waves.

17. 3/4 time music has 3 quarter notes per measure and 4/4 time music has 4 quarter notes per measure. The difference is the number of quarter notes per measure.

Chapter Six

Practice Set 6–1

1. $(3, -2)$ solves both equations: $3(3) + 4(-2) = 1$ and $3 - (-2) = 5$
3. $(-3, -1)$ solves both equations: $-3 - 4(-1) = 1$ and $5(-3) - 3(-1) = -12$
5. $(½, 1)$ solves both equations: $2(½) + 1 = 2$ and $6(½) - 4(1) = -1$

Note: for numbers 7–25, small graphs here would be of little use to show the actual crossing points in enough detail to be readable

7. $(1,2)$ using a graphing calculator
9. $(2, 0)$ using a graphing calculator
11. $(4, 1)$ using a graphing calculator
13. $(0,1)$ using a graphing calculator
15. infinite solutions – same line; using a graphing calculator
17. no solution – parallel lines; using a graphing calculator
19. $(-2,-4)$ using a graphing calculator
21. $(-3,-1)$ using a graphing calculator
23. infinite solutions – same line; using a graphing calculator
25. no solution – parallel lines; using a graphing calculator
27. $(2, 0)$ using a graphing calculator
29. $(2,-1)$ using a graphing calculator
31. infinite solutions – same line; using a graphing calculator

Practice Set 6–2

1. $x = 2y$; Substitute this for x in the second equation.

    $2y + y = 18$
    $3y = 18$
    $y = 6$

    Substitute $y = 6$ into the first equation: $x = 2y = 2(6) = 12$

    The solution is $(12, 6)$.

3. $x = 5y + 5$; Substitute this in for x in the first equation.

    $5(5y + 5) + 3y = 11$

$$28y + 25 = 11$$
$$28y = -14$$
$$y = -0.5$$

Substitute $y = -0.5$ into the second equation: $x = 5y + 5 = 5(-0.5) + 5 = 2.5$; The solution is $(2.5, -0.5)$.

5. Solve the first equation for y: $y = 3x - 4$ and substitute this in for y in the second equation.

$$9x - 5(3x - 4) = -10$$
$$9x - 15x + 20 = -10$$
$$-6x = -30$$
$$x = 5$$

Substitute $x = 5$ into the first equation: $3(5) - y = 4$

$$15 - y = 4$$
$$-y = -11$$
$$y = 11 \quad \text{The solution is } (5, 11).$$

7. $2x - 3y = 7$

 $4x + 3y = 5$ Multiply equation #1 by -2 and add to equation #2

$$-4x + 6y = -14$$
$$\underline{4x + 3y = 5}$$
$$9y = -9$$
$$y = -1$$

Substitute $y = -1$ into the first equation: $2x - 3(-1) = 7$

$$2x + 3 = 7$$
$$2x = 4$$
$$x = 2 \quad \text{The solution is } (2, -1).$$

9. $x + 7y = 12$

 $3x - 5y = 10$ Multiply equation #1 by -3 and add to equation #2

$$-3x - 21y = -36$$
$$\underline{3x - 5y = 10}$$
$$-26y = -26$$
$$y = 1$$

Substitute y = 1 into the first equation: x + 7(1) = 12

$$x + 7 = 12$$
$$x = 5 \quad \text{The solution is } (5, 1).$$

11. x − y = 3 (after rearrangement)

    x − 3y = −2 (after rearrangement)

    Multiply equation #1 by −1 and add to equation #2

    $$-x + y = -3$$
    $$\underline{x - 3y = -2}$$
    $$-2y = -5$$
    $$y = 2.5$$

    Substitute y = 2.5 into the first equation: x − 2.5 = 3

    $$x = 5.5 \quad \text{The solution is } (5.5, 2.5).$$

13. 3x − 2y = 6

    −6x + 4y = −12   Multiply equation #1 by 2 and add to equation #2

    $$6x - 4y = 12$$
    $$\underline{-6x + 4y = -12}$$
    $$0 = 0$$

    This means that these are the same line and there are an infinite number of solutions.

15. multiplying diagonals gives (1)(5) − (2)(2) = 5 − 4 = 1

17. multiplying diagonals gives (3)(4) − (6)(2) = 12 − 12 = 0

19. multiplying diagonals gives (3)(4) − (−2)(5) = 12 + 10 = 22

21. $x = \dfrac{\begin{vmatrix} 1 & -4 \\ 13 & 3 \end{vmatrix}}{\begin{vmatrix} 1 & -4 \\ 2 & 3 \end{vmatrix}} = \dfrac{3-(-52)}{3-(-8)} = \dfrac{55}{11} = 5$

    Now substitute into equation #1. $5 - 4y = 1$
    $$-4y = -4$$
    $$y = 1 \quad \text{The solution is } (5, 1)$$

23. $x = \dfrac{\begin{vmatrix} 36 & -2 \\ 47 & 4 \end{vmatrix}}{\begin{vmatrix} 6 & -2 \\ 5 & 4 \end{vmatrix}} = \dfrac{144-(-94)}{24-(-10)} = \dfrac{238}{34} = 7$

Now substitute into equation #1.

$6(7) - 2y = 36$

$42 - 2y = 36$

$-2y = -6$

$y = 3$ The solution is (7, 3).

25. $x = \dfrac{\begin{vmatrix} 28 & -7 \\ -20 & 5 \end{vmatrix}}{\begin{vmatrix} 6 & -7 \\ -4 & 5 \end{vmatrix}} = \dfrac{140 - 140}{30 - 28} = \dfrac{0}{2} = 0$

Now substitute into equation #1.

$6(0) - 7y = 28$

$-7y = 28$

$y = -4$ The solution is (0, −4).

27. $x = \dfrac{\begin{vmatrix} -10 & -1 \\ 4 & 1 \end{vmatrix}}{\begin{vmatrix} 2 & -1 \\ 1 & 1 \end{vmatrix}} = \dfrac{-10 - (-4)}{2 - (-1)} = \dfrac{-6}{3} = -2$

Now substitute into equation #2.

$-2 + y = 4$

$y = 6$ The solution is (−2, 6).

29. $x = \dfrac{\begin{vmatrix} 13 & -2 \\ -26 & 4 \end{vmatrix}}{\begin{vmatrix} 3 & -2 \\ -6 & 4 \end{vmatrix}} = \dfrac{52 - 52}{12 - 12} = \dfrac{0}{0} =$ infinite solutions, same line

31. $x = \dfrac{\begin{vmatrix} 0 & 1 \\ 86 & -15 \end{vmatrix}}{\begin{vmatrix} -3 & 1 \\ 2 & -15 \end{vmatrix}} = \dfrac{0 - 86}{45 - 2} = \dfrac{-86}{43} = -2$

Now substitute into equation #1.

$y = 3(-2)$

$y = -6$ The solution is (−2, −6).

33. $x = \dfrac{\begin{vmatrix} 10 & -3 \\ 2 & 5 \end{vmatrix}}{\begin{vmatrix} 2 & -3 \\ 2 & 5 \end{vmatrix}} = \dfrac{50-(-6)}{10-(-6)} = \dfrac{56}{16} = 3.5$

Now substitute into equation #1. $2(3.5) - 3y = 10$

$7 - 3y = 10$

$-3y = 3$

$y = -1$ The solution is (3.5, −1).

35. $x = \dfrac{\begin{vmatrix} 0 & 4 \\ 17 & -5 \end{vmatrix}}{\begin{vmatrix} 1 & 4 \\ 3 & -5 \end{vmatrix}} = \dfrac{0-68}{-5-12} = \dfrac{-68}{-17} = 4$

Now substitute into equation #1. $4 = -4y$

$-1 = y$ The solution is (4, −1).

37. $x = \dfrac{\begin{vmatrix} -36 & 3 \\ 22 & -5 \end{vmatrix}}{\begin{vmatrix} 2 & 3 \\ 3 & -5 \end{vmatrix}} = \dfrac{180-66}{-10-9} = \dfrac{114}{-19} = -6$

Now substitute into equation #1. $3y + 2(-6) = -36$

$3y - 12 = -36$

$3y = -24$

$y = -8$ The solution is (−6, −8).

39. $x = \dfrac{\begin{vmatrix} 2.2 & -0.4 \\ 1.6 & -1 \end{vmatrix}}{\begin{vmatrix} 0.2 & -0.4 \\ 0.5 & -1 \end{vmatrix}} = \dfrac{-2.2-(-0.64)}{-0.2-(-0.2)} = \dfrac{-1.56}{0} =$ no solution; parallel lines

41. $x = \dfrac{\begin{vmatrix} 1 & 3 \\ 2 & 2 \end{vmatrix}}{\begin{vmatrix} 2 & 3 \\ 3 & 2 \end{vmatrix}} = \dfrac{2-6}{4-9} = \dfrac{-4}{-5} = \dfrac{4}{5}$

Now substitute into equation #1.
$$2(\frac{4}{5}) + 3y = 1$$
$$\frac{8}{5} + 3y = 1$$
$$3y = -\frac{3}{5}$$
$$y = -\frac{1}{5} \quad \text{The solution is } \left(\frac{4}{5}, -\frac{1}{5}\right).$$

43. $x = \dfrac{\begin{vmatrix} 1 & -2/5 \\ 2 & -3/2 \end{vmatrix}}{\begin{vmatrix} 2/3 & -2/5 \\ 5/2 & -3/2 \end{vmatrix}} = \dfrac{-3/2 - (-4/5)}{-6/6 - (-10/10)} = \dfrac{-7/10}{0} =$ no solution; parallel lines

45. By substitution; solve equation #1 for "y"; $y = 2x - 2$
   Now substitute for "y" in equation #2; $4x + 3(2x-2) = 24$
   $$4x + 6x - 6 = 24$$
   $$10x = 30$$
   $$x = 3$$
   Now substitute into equation #1: $y = 2(3) - 2 = 4$. The solution is (3, 4).

47. Substitute for "y" from equation #1 into equation #2: $2x - 1 = x + 1$
   $$x = 2$$
   Now use this value in equation #1; $y = 2(2) - 1 = 3$. The solution is (2, 3)

49. Solve by addition (equation #1) $5x - 2y = 14$; (equation #2) $6x - 5y = 9$;
   Eliminate "y" by multiplying equation #1 by 5 and equation #2 by –2 and then adding.
   $$25x - 10y = 70$$
   $$\underline{-12x + 10y = -18}$$
   $$13x \qquad = 52$$
   $$x = 4$$
   Substitute into equation #1; $5(4) - 2y = 14$
   $$20 - 2y = 14$$

$$-2y = -6$$
$$y = 3 \qquad \text{The solution is (4, 3).}$$

51. Solve by substitution. Rearrange equation #1: $y = x - 8$.

    Now substitute into equation #2.
    $$3(x - 8) = x + 2$$
    $$3x - 24 = x + 2$$
    $$2x = 26$$
    $$x = 13$$

    Now substitute $x = 13$ into equation #1.
    $$y = 13 - 8 = 5. \qquad \text{The solution is (13, 5).}$$

53. Solve by substitution. Divide equation #1 by -2 giving $-7x + 3 = y$ or $y = -7x + 3$.

    Substitute into equation #2; $\quad 7x + (-7x + 3) = 3$
    $$7(0) + (-7)(0) + 3 = 3$$
    $$0 = 0$$

    The lines are the same line, so there infinite solutions.

55. Solve by addition. Multiply equation #1 by 4 and equation #2 by 3 and then add.
    $$28x + 12y = -32$$
    $$\underline{15x - 12y = -183}$$
    $$43x \quad\quad = -215$$
    $$x = -5$$

    Substitute into equation #1: $\quad 7(-5) + 3y = -8$
    $$-35 + 3y = -8$$
    $$3y = 27$$
    $$y = 9 \qquad \text{The solution is } (-5, 9).$$

57. No, because (2, 3) is not a solution for equation #1. $\quad 3(2) + 5 = 11 \neq 3$

59. $x + y = 20$

    $\underline{x - y = 2}$

    $2x \quad\quad = 22$

    $\quad\quad x = 11$

    Substitute into equation #1: $\quad 11 + y = 20$
    $$y = 9 \qquad \text{The numbers are 11 and 9.}$$

61. Set up two equations as follows: $x + y = 30$ and $2x = 3y$.

solve equation #1 for "x" and solve by substitution into equation #2: $x = -y + 30$.

$$2(-y + 30) = 3y$$
$$-2y + 60 = 3y$$
$$5y = 60$$
$$y = 12$$

Substitute into equation #1: $x = -12 + 30 = 18$. The numbers are 18 and 20.

63. Set up two equations: $\dfrac{x+y}{2} = x - y$ and $x + y = 64$.

Solve equation #2 for x: $x = 64 - y$.

Simplify equation #1 by multiplying both sides by 2.

$$x + y = 2x - 2y$$
$$3y = x$$

Now substitute "x" from equation #1:

$$3y = 64 - y$$
$$4y = 64$$
$$y = 16$$

Substitute into equation #2: $x = 64 - 16 = 48$. The numbers are 48 and 16.

65. Set up two equations: $3x = 4y$ and $x + y = 49$.

Solve equation #2 for "x": $x = 49 - y$.

Substitute into equation #1: $3(49 - y) = 4y$

$$147 - 3y = 4y$$
$$147 = 7y$$
$$y = 21$$

Substitute into equation #2: $x = 49 - 21 = 28$. The numbers are 28 and 21.

Practice Set 6-3

1. Let x = 10% solution and y = 25% solution.

   Set up two equations: $0.10x + 0.25y = 0.18(30)$ and $x + y = 30$.

   Solve equation #2 for "y" and substitute into equation #1: $y = 30 - x$.

   $$0.10x + 0.25(30 - x) = 0.18(30)$$

$$0.10x + 7.5 - 0.25x = 5.4$$
$$-0.15x = -2.1$$
$$x = 14$$

Solve for y: $y = 30 - 14 = 16$.

Thus use 14 mL of 10% and 16 mL of 25%.

3. Let x = one-bedroom units and y = two-bedroom units.

   Set up 2 equations: $x + y = 20$ and $\$650x + \$825y = \$14{,}400$.

   Solve equation #1 for "y" and substitute into equation #2: $y = 20 - x$.

   $$650x + 825(20 - x) = 14{,}400$$
   $$650x + 16{,}500 - 825x = 14{,}400$$
   $$-175x = -2100$$
   $$x = 12$$

   Solve for y: $y = 20 - 12 = 8$

   So there are 12 one-bedroom units and 8 two-bedroom units.

5. The equilibrium price occurs when $D(p) = S(p)$.

   $$30 - 3p = 4p - 5$$
   $$35 = 7p$$
   $$p = 5 \qquad \text{So, the equilibrium price is \$5.}$$

7. Let the cost of Model A = cost of Model B.

   $$950 + 32x = 1275 + 22x$$
   $$10x = 325$$
   $$x = 32.5.$$ The costs of Model A and Model B will be equal in 32.5 years.

   In fifteen years, $950 + 32(15) = 950 + 480 = \$1430$ for Model A

9. Let x = 60% silver and y = 40% silver.

   Set up two equations: $0.60x + 0.40y = 0.52(20)$ and $x + y = 20$.

   Solve equation #2 for "y" and substitute into equation #1; $y = 20 - x$.

   $$0.6x + 0.4(20 - x) = 10.4$$
   $$0.6x + 8 - 0.4x = 10.4$$
   $$0.2x = 2.4$$
   $$x = 12$$

Solve for y: $y = 20 - 12 = 8$.

Thus, use 12 g of 60% and 8g of 40%.

11. John's Rent-A-Car: $26 + 0.15x = \text{Cost}$

    Cookie's Rental Agency: $18 + 0.20x = \text{Cost}$

    Set the two costs equal to each other: $26 + 0.15x = 18 + 0.20x$

    $$8 = 0.05x$$
    $$x = 160$$

    You will need to drive 160 miles for the cost of the two agencies to be the same.

    Cost at John's: $26 + 0.15(200) = 26 + 30 = \$56$

    Cost at Cookie's: $18 + 0.2(200) = 18 + 40 = \$58$; So, John's is cheaper at 200mi.

13. Set the two plans equal to each other.

    $$200 + 0.05x = 0.15x$$
    $$200 = 0.1x$$
    $$x = \$2000 \text{ needed for the commissions to be equal}$$

    $200 + 0.05(1500) = 200 + 75 = \$275$

    $0.15(1500) = \$225$

    So, he would do better with the $200 plus 5% commission if he made $1500 in sales.

15. At equilibrium: $D(p) = S(p)$.

    $$800 + 3.2p = 1200 - 2.8p$$
    $$-6p = -400$$
    $$p = 66.67 \text{ so the price of the product where supply would equal demand is \$66.67.}$$

17. Set up two equations where x= pizzas and y= sodas.

    Math Club: $3x + 4y = 61$ and Physics Club: $2x + 3y = 42$

    Now multiply equation #1 by –2, multiply equation #2 by 3, and add to equation #2.

    $$-6x - 8y = -122$$
    $$\underline{6x + 9y = \ 126}$$
    $$y = 4$$

    Substitute to find x: $3x + 4(4) = 61$

    $$3x + 16 = 61$$
    $$3x = 45$$

$$x = 15$$

So a large pizza costs $15 and a pitcher of soda costs $4.

19. Let x = the length and y = the width.

$$x = 3 + y$$
$$2x + 2y = 14$$

Solve by substituting equation #1 into equation #2.

$$2(3 + y) + 2y = 14$$
$$6 + 2y + 2y = 14$$
$$6 + 4y = 14$$
$$4y = 8$$
$$y = 2$$

Calculate the length: x = 3 + 2 = 5, so the length is 5 ft and the width is 2ft.

21. Let x = candy selling for $3.40 and y = candy selling for $4.60.

$$x + y = 10$$
$$3.40x + 4.60y = 4(10)$$

Solve equation #1 for x and substitute into equation #2: x = 10 – y.

$$3.40(10 - y) + 4.60y = 40$$
$$34 - 3.4y + 4.60y = 40$$
$$34 + 1.2y = 40$$
$$1.2y = 6$$
$$y = 5$$

Solve for the value of x: x = 10 – 5 = 5, so he needs 5 pounds of each type of candy.

23. Income = 32x and Cost = 25,500 + 15x. Set them equal to each other.

$$32x = 25,500 + 15x$$
$$17x = 25,500$$
$$x = 1500$$

You need to sell 1500 backpacks to break even.

25. Let x = length and y = width.

$$x = 3y$$
$$2x + 2y = 296$$

Substitute equation #1 into equation #2 and solve for y.

$$2(3y) + 2y = 296$$

Fundamentals of Algebraic Modeling, 5th Ed.

$$6y + 2y = 296$$
$$8y = 296$$
$$y = 37$$

Solve for the length: $x = 3(37) = 111$.   So, the dimensions are 111ft by 37ft.

27. Let x = the cost of a double room and y = the cost of a single room.

$$90x + 80y = 6930$$
$$x + y = 80$$

Solve equation #2 for x and substitute into equation #1: $x = 80 - y$

$$90(80 - y) + 80y = 6930$$
$$7200 - 90y + 80y = 6930$$
$$7200 - 10y = 6930$$
$$-10y = -270$$
$$y = 27$$

Solve for x using this value: $x = 80 - 27 = 53$

So, there are 53 double rooms and 27 single rooms.

29. Let x = 10% solution and y = 60% solution.

$$0.10x + 0.60y = 0.40(50)$$
$$x + y = 50$$

Solve equation #2 for x and substitute into equation #1: $x = 50 - y$

$$0.10(50 - y) + 0.60y = 20$$
$$5 - 0.10y + 0.60y = 20$$
$$0.5y = 15$$
$$y = 30$$

Use this value to solve for x: $x = 50 - 30 = 20$

So, he should mix 20 mL of 10% solution with 30 mL of 60% solution.

31. Let x = length and y = width.

$$2x + 2y = 54$$
$$x = 2y - 3$$

Substitute equation #2 into equation #1.

$$2(2y - 3) + 2y = 54$$
$$4y - 6 + 2y = 54$$

$$6y - 6 = 54$$
$$6y = 60$$
$$y = 10$$

Use this value to solve for x: $x = 2(10) - 3 = 20 - 3 = 17$

The dimensions of the lawn are 10 m by 17 m.

33. Let x = two congruent sides of the triangle and y = third side.

$$2x + y = 48$$
$$y = x - 3$$

Substitute equation #2 into equation #1.

$$2x + (x - 3) = 48$$
$$3x - 3 = 48$$
$$3x = 51$$
$$x = 17$$

So, the lengths of the congruent sides are 17 inches and the base is 17 − 3 = 14 inches.

35. Let x = money in the 5% investment and y = money in the 4.5% investment.

$$x + y = 12{,}000$$
$$0.05x + 0.045y = 560$$

Solve equation #1 for x and substitute into equation 2: $x = 12{,}000 - y$

$$0.05(12{,}000 - y) + 0.045y = 560$$
$$600 - 0.05y + 0.045y = 560$$
$$600 - 0.005y = 560$$
$$-0.005y = -40$$
$$y = 8000$$

Use this value to calculate the value of x: $x = 12{,}000 - 8000 = 4000$

So, $8000 is invested at 4.5% and $4000 is invested at 5%.

Practice Set 6-4

1. $y = x^2$ and $3x + y = 10$

Substitute equation #1 into equation #2. Set it equal to zero and solve.

$$3x + x^2 = 10$$
$$x^2 + 3x - 10 = 0$$

Fundamentals of Algebraic Modeling, 5th Ed.

factor: $(x + 5)(x - 2) = 0$

$x = -5$ or 2

In equation #1: if $x = -5$, then $y = (-5)^2 = 25$; if $x = 2$, then $y = 2^2 = 4$.

So the solutions are $\{(2, 4), (-5, 25)\}$.

3. $y = x$ and $y = -2x^2 + 1$

Substitute equation #1 into equation #2. Set it equal to zero and factor to solve.

$$x = -2x^2 + 1$$
$$2x^2 + x - 1 = 0$$

factor: $(2x - 1)(x + 1) = 0$

$x = 0.5$ or -1

Since $y = x$, the solutions are $\{(-1, -1), (0.5, 0.5)\}$.

5. $y = x^2 - 5x + 6$ and $y = 2x$

Substitute equation #2 into equation #1. Set it equal to 0 and factor to solve.

$$2x = x^2 - 5x + 6$$
$$x^2 - 7x + 6 = 0$$

factor: $(x - 6)(x - 1) = 0$

$x = 6$ or 1

In equation #2: if $x = 6$, then $y = 2(6) = 12$; if $x = 1$, then $y = 2(1) = 2$.

The solutions are $\{(6, 12), (1, 2)\}$.

7. $y = x^2 - 4$ and $y = 2x - 1$

Substitute equation #2 into equation #1. Set it equal to zero and factor to solve.

$$2x - 1 = x^2 - 4$$
$$x^2 - 2x - 3 = 0$$

factor: $(x - 3)(x + 1) = 0$

$x = 3$ or -1

In equation #2: if $x = 3$, then $y = 2(3) - 1 = 5$; if $x = -1$, then $y = 2(-1) - 1 = -3$.

The solutions are $\{(3, 5), (-1, -3)\}$.

9. $2x + 3y^2 = 2$ and $x - y = 1$

Reorder equation #2: $x = y + 1$ and substitute into equation #1.

$$2(y + 1) + 3y^2 = 2$$
$$3y^2 + 2y + 2 = 2$$

$$3y^2 + 2y = 0$$
factor: $$y(3y + 2) = 0$$
$$y = 0 \text{ or } -2/3$$

In equation #2:

if $y = 0$, then $x - 0 = 1$ and $x = 1$; if $y = -2/3$, then $x - (-2/3) = 1$ and $x = 1/3$.

The solutions are $\{(1, 0), (1/3, -2/3)\}$.

11. Add the two equations as they are.

$$x^2 + y^2 = 13$$
$$\underline{x^2 - y^2 = 5}$$
$$2x^2 = 18$$
$$x^2 = 9$$
$$x = 3 \text{ or } -3$$

In equation #1: if $x = 3$, then $3^2 + y^2 = 13$
$$9 + y^2 = 13$$
$$y^2 = 4$$
$$y = 2 \text{ or } -2$$

In equation #1: if $x = -3$, the answers are also 2 or -2.

So, the solutions are $\{(3, 2)\,(3,-2)\,(-3,2)\,(-3,-2)\}$.

13. $x + y = 7$ and $x^2 + y^2 = 25$

Solve for "y" in equation #1 and substitute into equation #2: $y = 7 - x$.

$$x^2 + (7 - x)^2 = 25$$
$$x^2 + 49 - 14x + x^2 = 25$$
$$2x^2 - 14x + 24 = 0$$
$$x^2 - 7x + 12 = 0$$
factor: $$(x - 3)(x - 4) = 0$$
$$x = 3 \text{ or } 4$$

Using equation #1 ($x + y = 7$), if $x = 3$, then $y = 4$ and if $x = 4$, then $y = 3$.

The numbers are 3 and 4.

15. $x^2 + y^2 = 58$ and $x - y = 10$

Solve equation #2 for x and substitute into equation #1: $x = y + 10$.

$$(y + 10)^2 + y^2 = 58$$

$$y^2 + 20y + 100 + y^2 = 58$$
$$2y^2 + 20y + 42 = 0$$
$$y^2 + 10y + 21 = 0$$
factor: $(y + 3)(y + 7) = 0$
$$y = -3 \text{ or } -7$$

Similarly, solving for x will give x = 3 or 7 so, the solutions are 3 and –7 or 7 and –3.

17. $xy = 143 \text{ ft}^2$ and $2x + 2y = 48 \text{ft}$

Solve equation #2 for y: $2y = 48 - 2x$
$$y = 24 - x$$

Now substitute into equation #1: $x(24 - x) = 143$
$$24x - x^2 = 143$$
$$x^2 - 24x + 143 = 0$$
factor: $(x - 11)(x - 13) = 0$
$$x = 11 \text{ or } 13$$

The solution is 13 feet by 11 feet

19. $xy = 60$ and $y - x = 7$

Solve equation #2 for y and substitute into equation #1: $y = x + 7$.
$$x(x + 7) = 60$$
$$x^2 + 7x = 60$$
$$x^2 + 7x - 60 = 0$$
factor: $(x - 5)(x + 12) = 0$

x = 5 or –12 and since the numbers have to be positive, x = 5.

Now solve for y: $y - 5 = 7$, so y = 12 and the numbers are 12 and 5.

21. $-0.1x^2 - 2x + 90 = 0.1x^2 - x + 30$

Solve for x: $0.2x^2 + x - 60 = 0$

Make all coefficients whole numbers by multiplying by 5
$$x^2 + 5x - 300 = 0$$
factor $(x - 15)(x + 20) = 0$
$$x = 15 \text{ or } -20$$

The answer must be positive so choose x = 15, or 15,000 pairs of sunglasses.

Now solve for the equilibrium price in either equation.
$$p = 0.1(15)^2 - 15 + 30 = 0.1(225) + 15 = \$37.50 \text{ each.}$$

Chapter 6 Review Problems

1. Substitute into both equations: $0 - 2(-2) = 4$ giving $4 = 4$ and
$$2(0) - 2 = -2 \text{ giving } -2 = -2$$
   Yes, the pair solves both equations.

2. Substitute into both equations: $3(-5) - (-2) = -13$ giving $-13 = -13$ and
$$-5 - 2(-2) = -1 \text{ giving } -1 = -1$$
   Yes, the pair solves both equations.

3. Substitute into both equations: $1 + 3 = 4$ giving $4 = 4$ and
$$2(1) - 3 = 5 \text{ giving } -1 \neq 5$$
   No, this is not the solution of the system since it does not satisfy both equations.

4. Substitute into both equations: $3(2) + 1 = 7$ giving $7 = 7$ and
$$2 = 2(1) \text{ giving } 2 = 2$$
   Yes, the pair solves both equations.

Small sketch graphs are very inaccurate in a key like this. Your graph crossing points should closely match the points given here for problem numbers 5–9.

5. (7, 1)
6. no solution; parallel lines
7. (3, 2)
8. (1, 3)
9. (0, 2)
10. $3x - y = 7$
    $\underline{2x + y = 3}$
    $5x = 10$
    $x = 2$
    Substitute into equation #2: $2(2) + y = 3$
    $ 4 + y = 3$
    $ y = -1$ The solution is $(2, -1)$.

11. $8x - 5y = 32$
    $\underline{4x + 5y = 4}$
    $12x = 36$
    $x = 3$

Substitute into equation #2: $4(3) + 5y = 4$

$$12 + 5y = 4$$
$$5y = -8$$
$$y = -1.6 \qquad \text{The solution is } (3, -1.6).$$

12. Multiply equation #2 by 2 and add to equation #1.

    $8x + 2y = 2$
    $\underline{6x - 2y = 12}$
    $14x = 14$
    $x = 1$

    Substitute into equation #1: $8(1) + 2y = 2$

    $$2y = -6$$
    $$y = -3 \qquad \text{The solution is } (1, -3).$$

13. $3x + 5y = 69$ and $y = 4x$

    Substitute for "y" in equation #1: $3x + 5(4x) = 69$

    $$23x = 69$$
    $$x = 3$$

    Substitute this value into equation #2: $y = 4(3) = 12$ \qquad The solution is $(3, 12)$.

14. Multiply equation #1 by −3 and add it to equation #2.

    $-18x - 9y = 0$
    $\underline{18x + 9y = 0}$
    $0 = 0$

    These are the same line with an infinite number of solutions.

15. Multiply equation #2 by 2 and add to equation #1.

    $6x - 8y = 6$
    $\underline{-6x + 4y = -4}$
    $-4y = 2$
    $$y = -\frac{1}{2}$$

    Substitute this value into equation #1: $6x - 8(-\frac{1}{2}) = 6$

    $$6x + 4 = 6$$

$$6x = 2$$
$$x = \frac{2}{6} = \frac{1}{3} \quad \text{The solution is} \left(\frac{1}{3}, -\frac{1}{2}\right).$$

16. $2x + 10y = 3$ and $x = 1 - 5y$

    Substitute equation #2 into equation #1: $2(1 - 5y) + 10y = 3$
    $$2 - 10y + 10y = 0$$
    $$2 = 0 \quad \text{(false statement)}$$

    These lines are parallel, so there is no solution.

17. Multiply equation #2 by -4, rearrange and add to equation #1.
    $$4x + 5y = 44$$
    $$\underline{-4x + 8y = 8}$$
    $$13y = 52$$
    $$y = 4$$

    Substitute this value into equation #1: $4x + 5(4) = 44$
    $$4x + 20 = 44$$
    $$4x = 24$$
    $$x = 6 \qquad \text{The solution is } (6, 4).$$

18. Multiply equation #1 by 3 and equation #2 by -2 and then add
    $$6x + 9y = -15$$
    $$\underline{-6x - 8y = 16}$$
    $$y = 1$$

    Substitute into equation #1: $2x + 3(1) = -5$
    $$2x + 3 = -5$$
    $$2x = -8$$
    $$x = -4 \qquad \text{The solution is } (-4, 1).$$

19. $x = \dfrac{\begin{vmatrix} 4 & -1 \\ 21 & 1 \end{vmatrix}}{\begin{vmatrix} 2 & -1 \\ 3 & 1 \end{vmatrix}} = \dfrac{4 - (-21)}{2 - (-3)} = \dfrac{25}{5} = 5$

    Substitute to solve for y: $3(5) + y = 21$
    $$15 + y = 21$$
    $$y = 6 \qquad \text{The solution is } (5, 6).$$

20. $x = \dfrac{\begin{vmatrix} 2 & 2 \\ 7 & 4 \end{vmatrix}}{\begin{vmatrix} 4 & 2 \\ 5 & 4 \end{vmatrix}} = \dfrac{8-14}{16-10} = \dfrac{-6}{6} = -1$

Substitute to solve for y: $4(-1) + 2y = 2$

$$-4 + 2y = 2$$
$$2y = 6$$
$$y = 3$$

The solution is (–1,3).

21. $x = \dfrac{\begin{vmatrix} 2 & 5 \\ 1 & -10 \end{vmatrix}}{\begin{vmatrix} 4 & 5 \\ -8 & -10 \end{vmatrix}} = \dfrac{-20-5}{-40+40} = \dfrac{-25}{0} =$ undefined

These are parallel lines with no solution.

22. $x = \dfrac{\begin{vmatrix} -4 & 1 \\ 7 & -1 \end{vmatrix}}{\begin{vmatrix} 2 & 1 \\ 1 & -1 \end{vmatrix}} = \dfrac{4-7}{-2-1} = \dfrac{-3}{-3} = 1$;

Substitute to find y: $2(1) + y = -4$

$$2 + y = -4$$
$$y = -6$$

The solution is (1, –6).

23. $x = \dfrac{\begin{vmatrix} 0.3 & -0.1 \\ -0.5 & -0.2 \end{vmatrix}}{\begin{vmatrix} 0.8 & -0.1 \\ 0.5 & -0.2 \end{vmatrix}} = \dfrac{-0.06-0.05}{-0.16-(-0.05)} = \dfrac{-0.11}{-0.11} = 1$

Substitute to find y: $0.8(1) - 0.1y = 0.3$

$$0.8 - 0.1y = 0.3$$
$$-0.1y = -0.5$$
$$y = 5$$

The solution is (1, 5).

24. Algebraic process:

Solve equation #1 for "y" and substitute into equation #2: $y = 8 - 2x$.

$$5x - 2(8 - 2x) = -16$$
$$5x - 16 + 4x = -16$$

$$9x = 0$$
$$x = 0$$

Solve for the value of y: $y = 8 - 2(0) = 8$. The solution is $(0, 8)$.

By Cramer's rule: $x = \dfrac{\begin{vmatrix} 8 & 1 \\ -16 & -2 \end{vmatrix}}{\begin{vmatrix} 2 & 1 \\ 5 & -2 \end{vmatrix}} = \dfrac{-16+(16)}{-4-5} = \dfrac{0}{-9} = 0$

Substitute as above to solve for "y" and the solution is $(0,8)$ as before.

25. Substitute equation #2 into equation #1 and solve for x by factoring.

$$x^2 + 2x - 3 = 0$$
$$(x + 3)(x - 1) = 0$$
$$x = -3 \text{ or } 1$$

Substitute both values into equation #2 to get the values of y: $y = 2(-3) - 3 = -9$ and $y = 2(1) - 3 = -1$. Thus, the solutions are $\{(-3, -9), (1, -1)\}$.

26. Substitute equation #2 into equation #1 and solve by factoring.

$$2x^2 + 1 = 3x + 6$$
$$2x^2 - 3x - 5 = 0$$
$$(2x - 5)(x + 1) = 0$$
$$x = 2.5 \text{ or } -1$$

Substitute these values into equation #1 and solve for y.

$$y = 3(2.5) + 6 = 13.5 \text{ and } y = 3(-1) + 6 = 3$$

The solutions are $\{(2.5, 13.5), (-1, 3)\}$.

27. $4x + 6.50y = 318$ and $x + y = 62$

Solve the second equation for x and substitute it into the first equation. $x = 62 - y$.

$$4(62 - y) + 6.50y = 318$$
$$248 - 4y + 6.50y = 318$$
$$2.50y = 70$$
$$y = 28$$

Substitute this value of y to find x: $x = 62 - 28 = 34$

Tyler sold 34 boxes of donuts and 28 stadium cushions.

28. $0.40x + 0.70y = .50(120)$ and $x + y = 120$

Solve the second equation for x and substitute into the first equation. $x = 120 - y$

$$0.40(120 - y) + 0.70y = 0.50(120)$$
$$48 - 0.40y + 0.70y = 60$$
$$0.30y = 12$$
$$y = 40$$

Substitute this value to find x: $x = 120 - 40 = 80$.

We will need 80 mL of the 40% acid solution and 40 mL of the 70% acid solution.

29. Let cost = income.

$$20,000 + 25x = 125x$$
$$20,000 = 100x$$
$$x = 200$$

200 clocks must be made and sold in order for the company to break even.

30. Let $D(p) = S(p)$.

$$120 - p = 2p - 117$$
$$-3p = -237$$
$$p = 79.$$

The equilibrium price is $79.00.

31. $x + y = 60$ and $0.10x + .05y = 5.75$

Solve the first equation for x and substitute into the second equation: $x = 60 - y$.

$$0.10(60 - y) + 0.05y = 5.75$$
$$6 - 0.10y + 0.05y = 5.75$$
$$-0.05y = -0.25$$
$$y = 5$$

Substitute this value to find x: $x = 60 - 5 = 55$

Cheryl had 55 dimes and 5 nickels.

32. $10x + 4y = 6.16$

$4x + 8y = 6.88$

Multiply the first equation by –2 and add the resulting equations.

$$-20x - 8y = -12.32$$
$$\underline{4x + 8y = 6.88}$$
$$-16x = -5.44$$
$$x = 0.34$$

Substitute this value to find y: $10(0.34) + 4y = 6.16$
$$3.4 + 4y = 6.16$$
$$4y = 2.76$$
$$y = 0.69$$

Potatoes are $0.34 per pound and bananas are $0.69 per pound.

33. Let $3000 + 3x = 5.50x$
$$3000 = 2.50x$$
$$x = 1200$$

The result is 1200 packages of stationery.

34. $10x + 15y = 98.25$
$$x = 0.20 + y$$

Substitute the second equation into the first and solve for y.
$$10(0.20 + y) + 15y = 98.25$$
$$2 + 10y + 15y = 98.25$$
$$25y = 96.25$$
$$y = 3.85.$$

Substitute this value to find x: $x = 0.20 + 3.85 = 4.05.$

The premium gas costs $4.05 per gallon.

35. answers will vary

Chapter 6 Test

1. $2(-3) + 3(4) = -6 + 12 = 6$; $2(-3) + 4 = -6 + 4 = -2$.

   Therefore, $(-3, 4)$ is the solution to this system of equations.

2. $2(6) - (-1)(-5) = 12 - 5 = 7$

For problems 3–5, your graphs should approximate the crossing points given.

3. $(4, 1)$

4. $(2, -1)$

5. no solution – parallel lines

6. $y = 4 - 3x$

   $3x + y = 5$

   Substitute the first equation into the second and solve for x.

   $$3x + 4 - 3x = 5$$

   $$4 = 5 \quad \text{This means that the lines are parallel with no solution.}$$

7. $x - 2y = 4$

   $3x + 4y = 2$

   Solve the first equation for x and substitute into the second equation: $x = 4 + 2y$

   $$3(4 + 2y) + 4y = 2$$

   $$12 + 6y + 4y = 2$$

   $$10y = -10$$

   $$y = -1$$

   Substitute this value to find x: $x = 4 + 2(-1) = 4 - 2 = 2$ The solution is $(2, -1)$.

8. $4x + 3y = 15$

   $2x - 5y = 1$

   Multiply the second equation by –2 and add the resulting equations together.

   $$4x + 3y = 15$$
   $$\underline{-4x + 10y = -2}$$
   $$13y = 13$$
   $$y = 1$$

   Substitute this value to find x: $4x + 3(1) = 15$

   $$4x + 3 = 15$$
   $$4x = 12$$
   $$x = 3 \quad \text{The solution is } (3, 1).$$

9. $2x + 4y = 7$

   $5x - 3y = -2$

   Multiply the first equation by 3 and the second by 4, then add the resulting equations.

   $$6x + 12y = 21$$
   $$\underline{20x - 12y = -8}$$
   $$26x = 13$$

$$x = \frac{1}{2}$$

Substitute this value to find y: $2(\frac{1}{2}) + 4y = 7$

$$1 + 4y = 7$$
$$4y = 6$$
$$y = \frac{3}{2}$$

The solution is $\left(\frac{1}{2}, \frac{3}{2}\right)$.

10. $x = \dfrac{\begin{vmatrix} 13 & -7 \\ 7 & 5 \end{vmatrix}}{\begin{vmatrix} 3 & -7 \\ 6 & 5 \end{vmatrix}} = \dfrac{65-(-49)}{15-(-42)} = \dfrac{65+49}{15+42} = \dfrac{114}{57} = 2$

Substitute to find y: $3(2) - 7y = 13$

$$6 - 7y = 13$$
$$-7y = 7$$
$$y = -1$$

The solution is $(2, -1)$.

11. $x = \dfrac{\begin{vmatrix} 0.3 & -1.2 \\ 1.6 & 1 \end{vmatrix}}{\begin{vmatrix} 0.5 & -1.2 \\ 0.2 & 1 \end{vmatrix}} = \dfrac{0.3-(-1.92)}{0.5-(-0.24)} = \dfrac{0.3+1.92}{0.5+0.24} = \dfrac{2.22}{0.74} = 3$

Substitute to find y: $0.5(3) - 1.2y = 0.3$

$$1.5 - 1.2y = 0.3$$
$$-1.2y = -1.2$$
$$y = 1$$

The solution is $(3, 1)$.

12. $2x^2 - y = 0$

$3x - y = -2$

Solve the first equation for y and substitute into the second equation: $y = 2x^2$.

$$3x - 2x^2 = -2$$
$$2x^2 - 3x - 2 = 0$$
$$(2x + 1)(x - 2) = 0$$
$$x = -1/2 = -0.5 \text{ and } x = 2.$$

Substitute these values to find y: $y = 2(-0.5)^2 = 2(0.25) = 0.5$

$$\text{and } y = 2(2)^2 = 2(4) = 8$$

The solutions to this system of equations are $(-0.5, 0.5)$ and $(2, 8)$.

13. Substitute equation #2 into equation #1 and solve by factoring.

$$x + 2(x^2 + 3x + 2) = 4$$
$$2x^2 + 7x = 0$$
$$x(2x + 7) = 0$$
$$x = 0 \text{ or } x = -3.5$$

Substitute both values into equation #2: $y = 0^2 + 3(0) + 2 = 2$
$$\text{and } y = (-3.5)^2 + 3(-3.5) + 2 = 3.75$$

The solutions are $(0, 2)$ and $(-3.5, 3.75)$.

14. Let Cost = Income.
$$10,000 + 5x = 25x$$
$$10,000 = 20x$$
$$x = 500$$

500 calculators must be sold for the company to break even.

15. $80x + 100y = 6400$

$x + y = 65$

Solve the second equation for x and substitute into the first equation: $x = 65 - y$
$$80(65 - y) + 100y = 6400$$
$$5200 - 80y + 100y = 6400$$
$$20y = 1200$$
$$y = 60$$

There are 60 double rooms.

16. $0.03x + 0.07y = 0.045(10)$

$x + y = 10$

Solve the second equation for x and substitute into the first equation: $x = 10 - y$
$$0.03(10 - y) + 0.07y = 0.045(10)$$
$$0.3 - 0.03y + 0.07y = 0.45$$
$$0.04y = 0.15$$
$$y = 3.75$$

Substitute this value to find x: $x = 10 - 3.75 = 6.25$
Mix 6.25 gallons of the 3% and 3.75 gallons of the 7%.

17. $0.105x + 0.12y = 1380$

$x + y = 12{,}000$

Solve the second equation for x and substitute into the first equation: $x = 12{,}000 - y$

$$0.105(12{,}000 - y) + .12y = 1380$$

$$1260 - 0.105y + 0.12y = 1380$$

$$0.015y = 120$$

$$y = 8000$$

Substitute to find x: $x = 12{,}000 - 8000 = 4000$

The amounts invested are $4000 at 10.5% and $8000 at 12%.

18. $2L + 2W = 76$ and $L = 2W + 5$

Substitute equation #2 into equation #1 and solve for W.

$$2(2W + 5) + 2W = 76$$

$$4W + 10 + 2W = 76$$

$$6W = 66$$

$$W = 11$$

Now substitute into equation #2: $L = 2(11) + 5 = 27$

The length = 27.

19. Equilibrium point is where Demand = Supply.

$$89 - 7p = 29 + 3p$$

$$60 = 10p$$

$$p = \$6$$

20. Let Cost = Income.

$$x(\$6) + \$17000 = x(\$10)$$

$$6x + 17000 = 10x$$

$$17000 = 4x$$

$x = 4250$ leotards need to be sold

Chapter Seven

Practice Set 7-1

1. True
3. False
5. False
7. False
9. False
11. True
13. A = {x| x is a natural number less than 6}
15. {2,4,6,8,10,12,14,....,40,42,44,46,48}
17. finite
19. infinite
21. {1,2,3,4,5,6,8}
23. ø or { }
25. {1, 4}

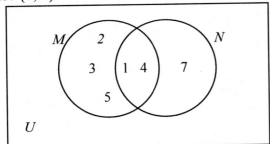

27. {1, 2, 3, 4, 5, 6, 8}

29. {2, 3, 5}

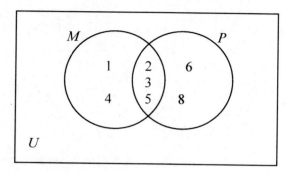

31. {1,2,3,4,5,7}

33. {1,2,3,5,7,9}

35. {3,5}

37. ∅ or { }

39. $\quad$ A ∩ (B ∩ C) =

$\quad\quad$ A ∩ {5, 7} =

{1, 2, 3, 4, 5} ∩ {5, 7} = {5}

Practice Set 7-2

1. Empirical probability is the relative frequency of the occurrence of an event. It is determined by actual observations of an experiment.

3. The law of large numbers states that the more times an experiment is performed, the more accurately we can predict the probability.

5. No. It means that the average person with traits similar to Mr. Timmons will live another 23.25 years.

7. Roll a die many times and find the relative frequency of odd numbers to the total number of rolls; ½; answers may vary.

9. $\dfrac{2350 \text{ females}}{4000 \text{ births}} = 58.75\%$; $\quad \dfrac{1650 \text{ males}}{4000 \text{ births}} = 41.25\%$

11. a. $\dfrac{64{,}528 \text{ males}}{81{,}800 \text{ couples}} = 78.89\% = 79\%$ $\quad$ b. $\dfrac{71{,}862 \text{ females}}{81{,}800 \text{ couples}} = 87.85\% = 88\%$

13. $\dfrac{12 \text{ blond hair}}{70 \text{ people}} = 17.14\%$

15. $\dfrac{265 \text{ made a C}}{672 \text{ people}} = 39.43\%$

17. $\dfrac{35 \text{ flounders}}{72 \text{ fish}} = 48.61\%$

19. 34 women are overweight (BMI ≥ 25.0); $\dfrac{34 \text{ BMI} \geq 25}{55 \text{ members}} = 61.8\%$

21. $\dfrac{52 \text{ BMI} \leq 34.9}{55 \text{ members}} = 94.5\%$

Practice Set 7-3

1. Outcomes that have the same chance of occurring in an experiment.

3. No. A theoretical probability of one means that the event is certain to occur.

5. There are 11 letters and 4 *s*'s. P(*s*) = $\dfrac{4}{11}$

7. There are no *a*'s so P(*a*) = 0.

9. $\dfrac{10 \text{ even numbers}}{20 \text{ numbers}} = \dfrac{1}{2}$

11. $\dfrac{1 \text{ three}}{20 \text{ numbers}} = \dfrac{1}{20}$

13. 1 − 0.375 = 0.625 or $\dfrac{625}{1000} = \dfrac{5}{8}$

15. $\dfrac{48 \text{ cards that are not three}}{52 \text{ cards in a deck}} = \dfrac{12}{13}$

17. $\dfrac{26 \text{ red}}{52 \text{ cards}} = \dfrac{1}{2}$

19. $\dfrac{6 \text{ black } K, Q, J}{52 \text{ cards}} = \dfrac{3}{26}$

21. $\dfrac{12 \text{ cards} > 5 \text{ but} < 9 \text{ (6,7,8 in each suit)}}{52 \text{ cards}} = \dfrac{3}{13}$

23. $\dfrac{7 \text{ marbles that are not blue}}{12 \text{ marbles}} = \dfrac{7}{12}$

25. $\dfrac{3\ even\ numbers}{6\ numbers\ on\ die} = \dfrac{1}{2}$

27. 0 (There are no numbers on the die greater than 6.)

29. Since the M&M's are produced with certain percentages of each color, the probability of drawing that particular color will match the composition. Therefore, P(*green*) =16% or $\dfrac{16}{100} = \dfrac{4}{25}$.

31. $\dfrac{1\ five\ on\ the\ die}{6\ possible\ outcomes} = \dfrac{1}{6}$

33. $\dfrac{2\ successes\ (either\ 1\ or\ 4\ on\ the\ die)}{6\ possible\ outcomes} = \dfrac{1}{3}$

35. $\dfrac{4\ successes\ (three\ odd\ numbers\ or\ a\ six\ on\ the\ die)}{6\ possible\ outcomes} = \dfrac{2}{3}$

37. $\dfrac{4\ aces}{52\ cards} = \dfrac{1}{13}$

39. $\dfrac{2\ cards\ that\ are\ K\ or\ Q\ of\ hearts}{52\ cards\ in\ a\ deck} = \dfrac{1}{26}$

41. $\dfrac{5\ odd\ numbers\ on\ the\ dial}{10\ possible\ channels} = \dfrac{1}{2}$

43. $\dfrac{1\ correct\ answer}{6\ possible\ answer\ choices} = \dfrac{1}{6}$

45. $\dfrac{35\ seconds\ red\ light}{90\ seconds\ for\ light\ to\ cycle\ through\ all\ colors} = \dfrac{7}{18}$

47. $\dfrac{55\ seconds\ yellow\ or\ green\ light}{90\ seconds\ for\ light\ to\ cycle\ through\ all\ colors} = \dfrac{11}{18}$

49. $\dfrac{24\ Duracell}{100\ batteries} = \dfrac{6}{25} = 0.24$

51. $\dfrac{84\ Duracell,\ Sony,\ or\ Eveready}{100\ batteries} = \dfrac{21}{25} = 0.84$

53. If the odds of winning are 3:17, that means 3 successes (or wins) and 17 failures (or losses) out of 20 events. The odds against winning are the same as the odds of losing which would be 17:3.

55. If there are 20 members of the class and 3 will be chosen, then 17 will not be chosen. Therefore, the odds of being chosen will be 3:17.

57. Since there are 3 even numbers and 3 odd numbers, the odds against rolling an odd number would be 3 even numbers : 3 odd numbers or 1:1.

59. There are 2 numbers less than 3 and 4 numbers that are greater than or equal to 3. Therefore the odds of rolling a number less than 3 are 2:4 or 1:2.

61. Success is drawing one of the four sixes. Failure is drawing a card that is not a six (48 cards). Therefore, the odds of success will be 4 sixes : 48 other cards or 1:12.

63. There are 12 face cards in the deck and 40 cards that are not face cards. The odds against drawing a face card are the same as the odds of drawing a non-face card. Therefore, the odds will be 40 non-face cards : 12 face cards or 10:3.

65. a. The odds against selecting a female are the same as the odds of selecting a male: 19 males:13 females or 19:13.
    b. The odds against selecting a male are the same as the odds of selecting a female: 13 females:19 males or 13:19.

67. If the probability of a mechanic fixing your car right the first time is 0.7 or $\frac{7}{10}$, then 7 times it is fixed right the first time and 3 times it isn't. Therefore, the odds against getting it fixed right the first time will be 3:7

69. Out of 1,000,000 raffle tickets, you have 10 tickets and there are 999,990 other tickets. Therefore, the odds of losing will be 999,990 other tickets chosen:10 tickets you bought or 99,999:1

71. If the odds are 5:9 in your favor, then there are 5 successes in 14 events. Therefore, P(*win*) = $\frac{5}{14}$.

73. If the odds against her being admitted are 9:2, then 9 times out of 11 she will not be admitted and 2 times out of 11 she will be admitted. Therefore, P(*no admission*) = $\frac{9}{11}$.

75. If the odds against promotion are 5:9, then, on average, 5 times in 14 he will not get a promotion and 9 times in 14 he will get a promotion. Therefore, P(*no promotion*) = $\frac{5}{14}$.

77. If the odds of winning are 7:5, that means 7 successes (wins) : 5 failures (losses). Since there are 12 total events, P(*loss*) = $\frac{5}{12}$.

79. If the odds of winning are 1:6, that means 1 success (win) : 6 failures (losses). Since there are 7 total events, P(*loss*) = $\frac{6}{7}$.

81. If the odds against winning are 5:2, that means 5 failures (losses) : 2 successes (wins). Since there are 7 total events, P(*loss*) = $\frac{5}{7}$.

Practice Set 7-4

1. 4 even numbers × 10 numbers × 10 numbers × 10 numbers = 4000 possible four-digit numbers
3. 5 bulbs in the bag × 5 bulbs in the bag × 5 bulbs in the bag = 125 different outcomes possible
5. 5 marbles × 4 marbles = 20 different outcomes possible
7. 2 sides (H,T) × 2 sides × 2 sides = 8 different outcomes possible
9. There are 8 possible outcomes with 3 outcomes consisting of one head and two tails: HTT, THT, TTH. Therefore, P(*one head and two tails*) = $\frac{3}{8}$.
11. There are two possible outcomes when tossing a penny {H, T} and six possible outcomes when rolling a die {1, 2, 3, 4, 5, 6}. Therefore, for the two events together there are 2 × 6 = 12 possible outcomes.
13. Only one of the 12 possible outcomes is H3, so P(*H3*) = $\frac{1}{12}$.
15.

1,1	1,2	1,3	1,4	1,5	1,6
2,1	2,2	2,3	2,4	2,5	2,6
3,1	3,2	3,3	3,4	3,5	3,6
4,1	4,2	4,3	4,4	4,5	4,6
5,1	5,2	5,3	5,4	5,5	5,6
6,1	6,2	6,3	6,4	6,5	6,6

   a. 36 possible outcomes
   b. There are six ways to roll the dice to obtain a total of 7: {1,6; 2,5; 3,4; 4,3; 5,2; 6,1}. Therefore, P(*sum of 7*) = $\frac{6}{36} = \frac{1}{6}$.
   c. 0; There are not two numbers whose sum will give 13.
   d. There is only one outcome where 3 is rolled on both dice, so P(*both 3*) = $\frac{1}{36}$.
17. {HH1, HH2, HH3, HH4, HH5, HH6, HT1, HT2, HT3, HT4, HT5, HT6, TH1, TH2, TH3, TH4, TH5, TH6, TT1, TT2, TT3, TT4, TT5, TT6}
19. The outcomes that include HT in any order and a 5 are HT5 and TH5. Therefore, P(*HT(in any order) and a 5*) = $\frac{2}{24} = \frac{1}{12}$.

21. 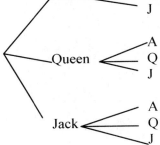   {AA, AQ, AJ, QA, QQ, QJ, JA, JQ, JJ}

23. There is only one outcome out of the 9 possible outcomes with JA. Therefore, $P(JA) = \dfrac{1}{9}$.

25. 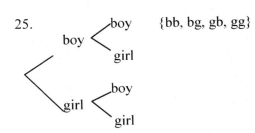   {bb, bg, gb, gg}

27. At least one boy includes all outcomes that have one or more boys: bg, gb, bb.

    Therefore, P(*at least one boy*) $= \dfrac{3}{4}$.

29. There are 26 letters in the alphabet and 10 numbers available (0-9). To calculate the number of possible passwords, multiply the number of outcomes for each event together.
    Letter, letter, letter, number, number gives $26 \times 26 \times 26 \times 10 \times 10 = 1{,}757{,}600$ possible passwords.

31. Possible combinations for three letters include 3 choices for first letter × 2 choices for second letter × 1 choice for third letter = 6 possible "words" using 3 letters.

Practice Set 7-5

1. $P(K \text{ or } Q) = \dfrac{4}{52} + \dfrac{4}{52} = \dfrac{8}{52} = \dfrac{2}{13}$; Yes, it is impossible to pick one card and get a king and a queen at the same time.

3. $P(\text{black or spade}) = \dfrac{26}{52} + \dfrac{13}{52} - \dfrac{13}{52} = \dfrac{26}{52} = \dfrac{1}{2}$; No, it is possible to pick one card and get a black spade because all spades are black!

5. $P(\text{face card or black card}) = \dfrac{12}{52} + \dfrac{26}{52} - \dfrac{6}{52} = \dfrac{32}{52} = \dfrac{8}{13}$; No; it is possible to pick one card and get a black face card. (jack of spades and clubs, queen of spades and clubs, king of spades and clubs).

Fundamentals of Algebraic Modeling, 5th Ed.

7. $P(2 \text{ or a number} > 2) = \dfrac{1}{8} + \dfrac{6}{8} = \dfrac{7}{8}$

9. $P(\text{odd or a number} > 2) = \dfrac{3}{6} + \dfrac{4}{6} - \dfrac{2}{6} = \dfrac{5}{6}$

11. The probability of drawing two face cards without replacement is the same as the probability of drawing one face card and then a second face card without putting the first card back into the deck.

    $P(\text{face card and a face card}) = \dfrac{12}{52} \times \dfrac{11}{51} = \dfrac{11}{221}$

13. Because the flowers are planted, the selections are done without replacement,

    $P(\text{white and white and white}) = \dfrac{12}{24} \times \dfrac{11}{23} \times \dfrac{10}{22} = \dfrac{5}{46} = 0.1087$

15. Because the flowers are planted, the selections are done without replacement,

    $P(\text{not red and not red and not red}) = \dfrac{17}{24} \times \dfrac{16}{23} \times \dfrac{15}{22} = \dfrac{85}{253} = 0.3360$

17. $P(\text{lose and win and win and lose and lose}) = \dfrac{3}{5} \times \dfrac{2}{5} \times \dfrac{2}{5} \times \dfrac{3}{5} \times \dfrac{3}{5} = \dfrac{108}{3125} = 0.03456$

19. $P(\text{lose and lose and lose and lose and lose}) = \dfrac{3}{5} \times \dfrac{3}{5} \times \dfrac{3}{5} \times \dfrac{3}{5} \times \dfrac{3}{5} = \left(\dfrac{3}{5}\right)^5 = \dfrac{243}{3125} = 0.07776$

21. There are 7 consonants and 4 vowels in the word *Mississippi*. Selection is done with replacement.

    a. $P(\text{consonant and consonant and consonant}) = \dfrac{7}{11} \times \dfrac{7}{11} \times \dfrac{7}{11} = \dfrac{343}{1331} = 0.2577$

    b. $P(\text{vowel and consonant and vowel}) = \dfrac{4}{11} \times \dfrac{7}{11} \times \dfrac{4}{11} = \dfrac{112}{1331} = 0.08415$

23. For a single birth, $P(\text{boy}) = \dfrac{1}{2}$ and $P(\text{girl}) = \dfrac{1}{2}$.

    For three births, $P(\text{a girl, then a girl, then a boy}) = \dfrac{1}{2} \times \dfrac{1}{2} \times \dfrac{1}{2} = \dfrac{1}{8} = 0.125$

25. For a single birth, $P(\text{girl}) = \dfrac{1}{2}$.

    For three births, $P(\text{three girls}) = \dfrac{1}{2} \times \dfrac{1}{2} \times \dfrac{1}{2} = \dfrac{1}{8} = 0.125$

27. There are 26 black cards in the deck and 13 of those are clubs.

    Therefore, $P(\text{club}|\text{black}) = \dfrac{13}{26} = 50\%$.

29. The sample space for a family with two children is {bb, bg, gb, gg}. Therefore, if you know the older child is a girl, $P(\text{two girls}|\text{the older of the children is a girl}) = \dfrac{1}{2}$.

31. a. $P(\text{good service}) = \dfrac{95 \text{ good service}}{145 \text{ total meals}} = \dfrac{19}{29} = 65.5\%$

    b. $P(\text{good service} | \text{meal was lunch}) = \dfrac{50 \text{ good service at lunch}}{65 \text{ lunch meals}} = \dfrac{10}{13} = 76.9\%$

    c. $P(\text{poor service} | \text{meal was dinner}) = \dfrac{35 \text{ poor service at dinner}}{80 \text{ dinner meals}} = \dfrac{7}{16} = 43.75\%$

    d. $P(\text{poor service} | \text{meal was lunch}) = \dfrac{15 \text{ poor service at lunch}}{65 \text{ lunch meals}} = \dfrac{3}{13} = 23.1\%$

33. a. $P(ABC \text{ or } NBC) = \dfrac{85}{230} + \dfrac{35}{230} = \dfrac{120}{230} = \dfrac{12}{23} = 52.2\%$

    b. $P(ABC | \text{woman}) = \dfrac{50 \text{ women watch ABC}}{95 \text{ women}} = \dfrac{10}{19} = 52.6\%$

    c. $P(ABC \text{ or } NBC | \text{man}) = \dfrac{35 \text{ men watch ABC and 25 men watch NBC}}{135 \text{ men}} = \dfrac{60}{135} = \dfrac{4}{9} = 44.4\%$

    d. $P(\text{not CBS} | \text{woman}) = \dfrac{75 \text{ women watch ABC or NBC or other}}{95 \text{ women}} = \dfrac{15}{19} = 78.9\%$

Practice Set 7-6

1. $_7C_2 = \dfrac{7!}{2!5!} = \dfrac{7 \cdot 6 \cdot 5 \cdot 4 \cdot 3 \cdot 2 \cdot 1}{(2 \cdot 1)(5 \cdot 4 \cdot 3 \cdot 2 \cdot 1)} = 21$

3. $_7P_2 = \dfrac{7!}{5!} = \dfrac{7 \cdot 6 \cdot 5 \cdot 4 \cdot 3 \cdot 2 \cdot 1}{(5 \cdot 4 \cdot 3 \cdot 2 \cdot 1)} = 42$

5. $9! = 9 \cdot 8 \cdot 7 \cdot 6 \cdot 5 \cdot 4 \cdot 3 \cdot 2 \cdot 1 = 362,880$

7. Only positive whole numbers have a factorial value. Thus, there is no value here.

9. In a combination, order is unimportant so that 1,2 and 2,1 are the same combination of the numbers 1 and 2. In a permutation, order is important so that 1,2 and 2,1 are different permutations (orderings) of the numbers 1 and 2.

11. If Jane must be on the team, then the other three members will be chosen from the nine remaining girls, and order doesn't matter here. $_9C_3 = 84$.

13. Because order is important when batting, this is a permutation. You can calculate it using the counting principle or a permutation. $9! = 362,880$ or $_9P_9 = 362,880$.

15. A person may choose no toppings, or 1 or 2 or...etc. So, calculate all possibilities and add.

    no toppings + 1 topping + 2 toppings + 3 toppings + 4 toppings + 5 toppings
    $$1 + {}_5C_1 + {}_5C_2 + {}_5C_3 + {}_5C_4 + {}_5C_5 = x$$
    $$1 + 5 + 10 + 10 + 5 + 1 = 32$$

17. Order is important in this problem, so a permutation can be used. You can also use the counting principle. $4! = 24$ or $_4P_4 = 24$.

19. a. Because order is important in a word, this is a permutation. You can calculate it using the counting principle or a permutation. $9! = 362,880$ or $_9P_9 = 362,880$

    b. Choosing 4 letters out of the nine to create different words by changing their order results in $_9P_4 = 3024$ different words.

21. a. Because order is important in a ship's message, this is a permutation. You can calculate it using the counting principle or a permutation. $8! = 40,320$ or $_8P_8 = 40,320$.

    b. Because order is important in a ship's message, this is a permutation. You can calculate it using the counting principle or a permutation. $8 \cdot 7 \cdot 6 = 336$ or $_8P_3 = 336$.

    c. Because order is important in a ship's message, this is a permutation. You can calculate it using the counting principle or a permutation. $8 \cdot 7 \cdot 6 \cdot 5 = 1680$ or $_8P_4 = 1680$.

23. Use the counting principle to calculate the total number of possible outcomes.

    8 entries · 9 entries = 72 daily-double tickets to purchase to guarantee a win.

25. Since any one of 10 numbers can be repeated in a social security number having 9 digits, there are $10^9 = 1,000,000,000$ possible social security numbers. (Remember that a permutation cannot be used if repetition of digits is allowed, so the counting principle must be used here.)

27. Since order is not important in choosing cars, a combination will give the answer. $_{10}C_6 = 210$.

29. Since order is not important in choosing finalists, a combination will give the answer. $_9C_3 = 84$.

31. You are choosing 3 marbles out of 9 and you want to find the probability that all are red.

    $$P(3\,red) = \frac{\text{choosing 3 red out of 5}}{\text{choosing 3 out of 9}} = \frac{{}_5C_3}{{}_9C_3} = \frac{10}{84} = \frac{5}{42} \text{ or } P(3\,red) = \frac{5}{9} \times \frac{4}{8} \times \frac{3}{7} = \frac{5}{42}.$$

33. You are choosing 2 numbers out of 6 and you want to find the probability that both are even.

    $$P(2\,even) = \frac{\text{choosing 2 even out of 3}}{\text{choosing 2 out of 6}} = \frac{{}_3C_2}{{}_6C_2} = \frac{3}{15} = \frac{1}{5} \text{ or } P(2\,even) = \frac{3}{6} \times \frac{2}{5} = \frac{1}{5}.$$

35. You are choosing 3 batteries out of 8 and you want to find the probability that all are good.

    $$P(3\,good) = \frac{\text{choosing 3 good out of 4}}{\text{choosing 3 out of 8}} = \frac{{}_4C_3}{{}_8C_3} = \frac{4}{56} = \frac{1}{14} \text{ or } P(3\,good) = \frac{4}{8} \times \frac{3}{7} \times \frac{2}{6} = \frac{1}{14}.$$

37. You are choosing 2 bills out of 8 and you want to find the probability that both are $5 bills.

$$P(2\ \$5\ bills) = \frac{choosing\ 2\ \$5\ bills\ out\ of\ 4}{choosing\ 2\ out\ of\ 8} = \frac{{}_4C_2}{{}_8C_2} = \frac{6}{28} = \frac{3}{14}\ \text{or}\ P(2\ \$5\ bills) = \frac{4}{8} \times \frac{3}{7} = \frac{3}{14}.$$

39. a. ${}_6P_6 = 720$ or using the counting principle: $6 \cdot 5 \cdot 4 \cdot 3 \cdot 2 \cdot 1 = 720$

    b. ${}_6P_4 = 360$ or using the counting principle: $6 \cdot 5 \cdot 4 \cdot 3 = 360$

41. $P(aisle\ seats) = \dfrac{get\ 4\ aisle\ seats\ out\ of\ 6}{get\ 4\ seats\ out\ of\ 10\ empty\ seats} = \dfrac{{}_6C_4}{{}_{10}C_4} = \dfrac{15}{210} = \dfrac{1}{14}$ or $P(aisle\ seats) =$

    $\dfrac{6}{10} \times \dfrac{5}{9} \times \dfrac{4}{8} \times \dfrac{3}{7} = \dfrac{1}{14}.$

43. a. The probability that the lot will pass is based on the probability of choosing 3 good chips from the 98 that are not defective. $P(3\ good\ chips) = \dfrac{choosing\ 3\ good\ chips\ out\ of\ 98}{choosing\ 3\ out\ of\ 100} = \dfrac{{}_{98}C_3}{{}_{100}C_3} = 0.940606$

    $= 94.1\%$ or $P(3\ good\ chips) = \dfrac{98}{100} \times \dfrac{97}{99} \times \dfrac{96}{98} = 0.940606 = 94.1\%$

    b. answers will vary

45. $P(uncle\ arrives\ first) = \dfrac{2\ uncles}{10\ family\ members} = \dfrac{1}{5}$

47. $P(four\ instructors) = \dfrac{choose\ 4\ out\ of\ 8\ instructors}{choose\ 4\ out\ of\ 15} = \dfrac{{}_8C_4}{{}_{15}C_4} = \dfrac{70}{1365} = \dfrac{2}{39} = 0.05128$

49. $P(3\ numbers > 4) = \dfrac{choosing\ 3\ numbers > 4\ out\ of\ 5\ possible}{choosing\ 3\ out\ of\ 10} = \dfrac{{}_5C_3}{{}_{10}C_3} = \dfrac{10}{120} = \dfrac{1}{12}$ or

    or $P(3\ numbers > 4) = \dfrac{5}{10} \times \dfrac{4}{9} \times \dfrac{3}{8} = \dfrac{1}{12}.$

51. There are 4 digits beginning with a five followed by three digits in each extension. Therefore, there are 10 choices for each of the three random digits so, using the counting principle, the total number of possible extensions is $10^3 = 1000$.

53. various answers

Fundamentals of Algebraic Modeling, 5th Ed.

Chapter 7 Review Problems

1.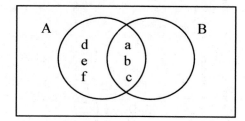

2. {a,b,c,e,g}

3. {f}

4. { } or ø

5. {a,c,e,f,g,h}

6. No. All elements in A are not contained in B.

7. No. All elements in C are not contained in A.

8. Yes; there are six elements in set A so it is finite.

9. No; there are two elements in set D so it is finite.

10. A probability is a mathematical calculation of the likelihood of a given event occurring in preference to all other possible events that could occur in a given situation (or experiment). In other words, it is the proportion (or fraction) of times that a particular outcome will occur.

11. In problems where several outcomes are possible, use the addition rule if the word "or" is used to connect the events and use the multiplication rule if the word "and" is used.

12. A probability is the proportion of outcomes considered favorable compared to the total number of possible results. The odds of an event compares outcomes considered favorable to the number of outcomes considered unfavorable.

13. There are 5 combinations that will total 8: {2,6} {3,5} {4,4} {5,3} {6,2}. So $P(sum\ of\ 8) = \dfrac{5}{36}$.

14. 0 (There are no combinations that will total 13.)

15. There are 15 combinations that give sums less than 7: {1,5} {1,4} {1,3} {1,2} {1,1} {2,1} {2,2} {2,3} {2,4} {3,1} {3,2} {3,3} {4,1} {4,2} {5,1}. $P(sum < 7) = \dfrac{15}{36} = \dfrac{5}{12}$.

16. $P(red) = \dfrac{7}{20}$

17. $P(green\ or\ blue) = \dfrac{8}{20} + \dfrac{5}{20} = \dfrac{13}{20}$.

18. $P(\text{three blue}) = \dfrac{5}{20} \times \dfrac{4}{19} \times \dfrac{3}{18} = \dfrac{1}{114}.$

19. $P(\text{male who drinks beer}) = \dfrac{63 \text{ male drinkers}}{200 \text{ students}} = \dfrac{63}{200} = 31.5\%$

20. $P(\text{female nondrinker}) = \dfrac{48 \text{ female nondrinkers}}{200 \text{ students}} = \dfrac{6}{25} = 24\%$

21. $P(\text{drinker} \mid \text{male student}) = \dfrac{63 \text{ male drinkers}}{113 \text{ males}} = \dfrac{63}{113} = 55.8\%$

22. $P(\text{woman} \mid \text{nondrinker}) = \dfrac{48 \text{ women nondrinkers}}{98 \text{ nondrinkers}} = \dfrac{24}{49} = 49.0\%$

23. There are three face cards (K, Q, J) per suit (4 suits); thus, there are 12 face cards. This means that there are 40 nonface cards in the deck. $s = 12, f = 40$.
    Odds in favor of drawing a face card = 12:40 = 3:10

24. $10^9 = 1$ billion

25. Yes, because by 2050 the population will be well over one billion.

26. $P(\text{good part}) = \dfrac{73}{75}$ or about a 97% chance of choosing a good one out.

27. $P(\text{rainy day}) = \dfrac{92}{365}$ or about a 25% chance of a rainy day.

28. $P(\text{ABC or NBC}) = \dfrac{80}{215} + \dfrac{30}{215} = \dfrac{110}{215} = \dfrac{22}{43} = 51.2\%$

29.
$P(\text{ABC or a woman}) = \dfrac{80 \text{ watch ABC}}{215} + \dfrac{100 \text{ women}}{215} - \dfrac{50 \text{ women watch ABC}}{215} = \dfrac{130}{215} = \dfrac{26}{43} = 60.5\%$

30. $P(\text{watches Fox} \mid \text{man}) = \dfrac{25 \text{ men watch Fox}}{115 \text{ men}} = \dfrac{5}{23} = 21.7\%$

31. $_7P_3 = 210$ or use the counting principle: $7 \times 6 \times 5 = 210$.

32. Two teachers and three students gives $_6C_2 \cdot {}_{50}C_3 = 294{,}000$ possible committees.

33. Since order is not important in choosing flavors, a combination will give the answer. $_{15}C_3 = 455$.

34. He has 8 possible successful tickets and there are 1492 other tickets. His odds of winning are 8:1492 or 2:373.

35. $4! = 4 \times 3 \times 2 \times 1 = 24$

36. Factorials do not apply to negative numbers. The answer is undefined.

37. Factorials do not apply to decimals. The answer is undefined.

38. $_7C_3 = \dfrac{7!}{3!4!} = \dfrac{7 \cdot 6 \cdot 5 \cdot 4 \cdot 3 \cdot 2 \cdot 1}{(3 \cdot 2 \cdot 1)(4 \cdot 3 \cdot 2 \cdot 1)} = 35$

39. $_7P_5 = \dfrac{7!}{2!} = \dfrac{7 \cdot 6 \cdot 5 \cdot 4 \cdot 3 \cdot 2 \cdot 1}{(2 \cdot 1)} = 2520$

40. $_{23}C_3 = \dfrac{23!}{3!20!} = \dfrac{23 \cdot 22 \cdot 21 \cdot 20 \cdot \ldots \cdot 3 \cdot 2 \cdot 1}{(3 \cdot 2 \cdot 1)(20 \cdot 19 \cdot 18 \cdot \ldots \cdot 2 \cdot 1)} = 1771$

Chapter 7 Test

1.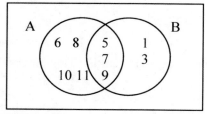

2. {5,7,9}

3. { } or ø

4. {10,11,12,13,15}

5. S = {H1, H2, H3, H4, H5, H6, T1, T2, T3, T4, T5, T6}

6. $P(\text{head and 5 on die}) = \dfrac{1}{2} \times \dfrac{1}{6} = \dfrac{1}{12}$.

7. $P(\text{tail and even}) = \dfrac{1}{2} \times \dfrac{3}{6} = \dfrac{1}{4}$.

8. $P(\text{adequate fire protection}) = \dfrac{85 \text{ adequate protection}}{150 \text{ residents}} = \dfrac{17}{30} = 56.7\%$

9. $P(\text{adequate} \mid \text{county resident}) = \dfrac{29 \text{ county who believe adequate}}{60 \text{ county residents}} = \dfrac{29}{60} = 48.3\%$

10. $P(\text{inadequate} \mid \text{city resident}) = \dfrac{34 \text{ city who believe inadequate}}{90 \text{ city residents}} = \dfrac{17}{45} = 37.8\%$

11. Using the counting principle, three numbers and three letters with no replacement gives $8 \times 9 \times 8 \times 26 \times 25 \times 24 = 8{,}985{,}600$ possible codes.

12. s = even number (3 on the die) and f = odd number (3 on the die). The odds are 3:3 = 1:1.

13. $P(\text{even number}) = \dfrac{3 \text{ even numbers}}{6 \text{ possible outcomes}} = \dfrac{1}{2}$.

14. $P(\text{Doris is promoted}) = 1 - 0.25 - 0.34 = 0.41$.

15. Using the counting principle, $3 \times 2 \times 1 = 6$.

16. 1 (oat)

17. $P(\text{meaningful word}) = \dfrac{1}{6}$.

18. Only nine plates out of all possible plates will have all four numbers the same.
    There are $9 \times 9 \times 9 \times 9 = 6561$ different possible plates. $P(4 \text{ numbers alike}) = \dfrac{9}{6561} = \dfrac{1}{729}$.

19. $P(red) = \dfrac{26 \text{ red cards}}{52 \text{ cards}} = \dfrac{1}{2}$.

20. $P(queen) = \dfrac{4 \text{ queens}}{52 \text{ cards}} = \dfrac{1}{13}$.

21. $P(\text{drawing 3 of diamonds twice}) = \dfrac{1}{52} \times \dfrac{1}{52} = \dfrac{1}{2704}$.

22. $P(\text{face card or diamond}) = \dfrac{12}{52} + \dfrac{13}{52} - \dfrac{3}{52} = \dfrac{22}{52} = \dfrac{11}{26}$.

23. $P(\text{Ringo wins}) = \dfrac{1}{4}$ $P(\text{Bonnie wins}) = \dfrac{1}{16}$

    $P(\text{Clyde wins}) = \dfrac{5}{13}$ $P(\text{Zero wins}) = \dfrac{2}{5}$

24. There are only two choices for children, boy or girl, each time, so for four births, $2 \times 2 \times 2 \times 2 = 16$ possible combination of boys and girls. But because order is not specified, many of the groupings are the same. If you list all possible combinations, 6 of the 16 will contain two boys and two girls. $P(\text{family with two boys and two girls}) = \dfrac{6}{16} = \dfrac{3}{8}$.

25. Order is important in this problem, so a permutation can be used. You can also use the counting principle. $6! = 720$ or $_6P_6 = 720$.

26. 0

27. $P(\text{missing a question}) = \dfrac{1}{2}$; $P(\text{missing 10 questions}) = \left(\dfrac{1}{2}\right)^{10} = \dfrac{1}{1024}$.

28. $P(\text{no improvement}) = \dfrac{24 \text{ no improvement}}{60 \text{ couples}} = \dfrac{2}{5} = 40\%$

29. $P(\text{couple with children} \mid \text{couple reported improvement}) = \dfrac{27}{36} = \dfrac{3}{4} = 75\%$

30.
    $P(\text{couple reported improvement \& has no children}) = \dfrac{9 \text{ reported improvement \& have children}}{60 \text{ couples}} = \dfrac{3}{20} = 15\%$

Fundamentals of Algebraic Modeling, 5th Ed.

Chapter Eight

Practice Set 8-1

1. Inferential; The results are based on a survey.
3. Descriptive; Though it may not be exact, it is still descriptive of the grades over a three year period.
5. Descriptive; Exact calculations are done after every major league game using the population of data.
7. Descriptive: This data is easily available in college records and is based on a complete population of data.
9. Descriptive; This number is based on gate counts for all attendees at each game.
11. numerical data; quantitative
13. numerical data; quantitative
15. non-numerical data; qualitative
17. Numbers are used but they represent personal preferences which is non-numerical data; qualitative
19. numerical data; quantitative
21. High schools were chosen based on geographical location, so this is a cluster sample.
23. Each student has an equal likelihood of being chosen, so it is simple random sample.
25. This survey is easy to do with no real plan or sequence and students are readily available, so this is a convenience sample.
27. Since two classes are chosen from all classes being taught by the instructor and all students in those classes are surveyed, this is a cluster sample.
29. All members have an equal chance of being chosen, so this is a simple random sample.
31. The numbering scale on the vertical axis exaggerates the differences.
33. Ask how many dentists were polled.
35. The number seems too precise for a general statement.
37. There is a misuse of a percent with a number greater than 100%.
39. The survey is not representative of the general population, just senior citizens.

Practice Set 8-2

1. Various answers will be given but should reflect an understanding of the meaning of "average".
3. sum = 46, n = 10, numbers sorted for finding the median and mode: 2, 2, 2, 3, 3, 4, 6, 7, 8, 9; mean = 4.6, median = 3.5, mode = 2; The best average is the mean since the mean and median are close to the same value.
5. sum = 251, n = 20, sorted list: 2,2,3,3,3,4,4,5,5,6,6,7,8,8,9,9,9,9,72,77 mean = 12.6, median = 6, mode = 9, The best average is the median since the mean and the median vary greatly due to the two very large values in the list.
7. sum = 43.79, n = 12, sorted list: 0.35, 1.23, 1.54, 2.65, 2.69, 2.85, 2.88, 3.21, 3.27, 4.25, 6.88, 11.99; mean = 3.65, median = 2.865, no mode; The best average is the median since the mean and the median vary greatly due to the one large value in the list.

9. A few persons in the US, like Bill Gates, Oprah Winfrey, and others make so much money in a year that they warp the mean average income for the entire nation.

11. a. the mode since you would like to see where "most" family sizes fall
    b. the mean because you want to be sure to include everyone who is a student
    c. the median was probably used to avoid having the average price warped by a few multimillion dollar homes
    d. the mean because the national debt would be divided by the total number of Americans (or adults in America)

13. $\bar{x} = \dfrac{2+5+3+0+0+(-3)+(-7)+1}{8} = \dfrac{1}{8} = 0.125°\text{C}$; median $= \tilde{x} = \dfrac{0+1}{2} = 0.5°\text{C}$
    (sum of the middle two numbers divided by 2); mode = 0° C

15. $\bar{x} = \dfrac{(4 \times 4)+(3 \times 3)+(3 \times 3)}{10} = \dfrac{34}{10} = 3.4$
    GPA = 3.4

17. Various explanations are possible, but the value is probably a median.

19. $\dfrac{(77.6 \times 5) + x}{6} = 83.0$
    $(77.6 \times 5) + x = 6(83.0)$
    $388 + x = 498$
    $x = 110$ (Based on a 100 point scale, she cannot attain an average of 83.)

21. a. $\bar{x} = \dfrac{1+2+3+4+4+7+11}{7} = \dfrac{32}{7} = 4.57$
    mean = 4.57 median = 4 mode = 4
    b. $\bar{x} = \dfrac{1+2+3+4+4+10+11}{7} = \dfrac{35}{7} = 5$
    The mean becomes 5 and the others remain unchanged.

23. original salaries: $\bar{x} = \dfrac{\$25{,}000 + \$30{,}000 + \$35{,}000}{3} = \dfrac{\$90{,}000}{3} = \$30{,}000$
    after the raise: $\bar{x} = \dfrac{\$27{,}000 + \$32{,}000 + \$37{,}000}{3} = \dfrac{\$96{,}000}{3} = \$32{,}000$
    The mean would increase by $2000.

25. No because the one extremely large salary will skew the mean to a higher number than the typical resident makes.

27. $\dfrac{x}{5} = 8$ where x = sum of five numbers
    $x = 40$

    $\dfrac{40 - y}{4} = 7$ where y = number removed from list
    $40 - y = 7(4)$
    $40 - y = 28$
    $40 - 28 = y$ The number that was removed is 12.

29. This requires that a class survey be done to determine the mode, mean and median.

Practice Set 8-3

1. sort by numbers less than 10, 10's, 20's, etc

   ```
   0 | 3 4 4 5 6 8 8 9
   1 | 0 0 1 1 2 2 2 2 2 3 4 5 5 6 6
   2 | 1 2 5 5 9
   3 | 1 6
   ```

3. a.
Classes	Frequency
50-54	2
45-49	0
40-44	2
35-39	2
30-34	3
25-29	11
20-24	5
15-19	13
10-14	7

   b.
   Trade Winds in Hawaii

   This is a bimodal distribution because there are two distinct peaks with a significant dip between them.

5. a. 3% are from Africa.
   b. Yes, because 36% > 33.3% or $\frac{1}{3}$.
   c. 49% of the total = (0.49)(7,500,000) = 3,675,000
   d. Oceania has the lowest percentage based on the graph.

7. a.
Classes	Frequency
1-5	8
6-10	11
11-15	13
16-20	11
21-25	4
26-30	4
31-35	2
36-40	3
41-45	2
46-50	2

b.

The graph is skewed right with lower frequencies occurring among higher commuting distances.

9. a. | Classes | Frequency |
   |---|---|
   | 70-74 | 8 |
   | 75-79 | 15 |
   | 80-84 | 8 |

b.

11. Various graphs are possible; this is a stem and leaf plot but a bar chart or histogram will do as well. We chose to sort by teens, 20's, etc. The student population spread shows that most students are in their teens and twenties.

```
1 | 8,8,9,9,9
2 | 0,0,2,2,2,3,3,5,7
3 |
4 | 2,3
5 | 7
```

Practice Set 8-4

1. All data items are identical.

3. $\bar{x} = \dfrac{0+1+2+3+4+5+6+7+8+9}{10} = \dfrac{45}{10} = 4.5$, $R = 9 - 0 = 9$, $s = 3.0$

5. $\bar{x} = \dfrac{0+0+2+5+6+9+11+15+18+18}{10} = \dfrac{84}{10} = 8.4$, $R = 18 - 0 = 18$, $s = 6.9$

7. $\bar{x} = \dfrac{6+6+10+12+3+5}{6} = \dfrac{42}{6} = 7$, $R = 12 - 3 = 9$, $s = 3.3$

9. $\bar{x} = \dfrac{4+0+3+6+9+12+2+3+4+7}{10} = \dfrac{50}{10} = 5$, $R = 12 - 0 = 12$, $s = 3.6$

11. Section 1: $R = 90 - 10 = 80$; $s = 24$
    Section 2: $R = 100 - 20 = 80$; $s = 27$
    Equal values for R might lead you to conclude that the sets were identical but, the different standard deviation values tell you that the sets are really different.

13. The smaller standard deviation for the Forever batteries (5 hours) indicates that their lifetime is much more likely to be close to the average of 155 hours. The larger standard deviation for the Enduring brand (20 hours) indicates that their lifetimes will vary more from the mean of 155 hours.

15. A small standard deviation is desirable and would indicate a consistent packing process and therefore, more consistency in the amount of chips in each bag.

17. a. Original Salaries: $\bar{x} = \dfrac{20{,}000 + 25{,}000 + 30{,}000}{3} = \dfrac{75{,}000}{3} = \$25{,}000$; $s = 5{,}000$

    With \$1000 Raise: $\bar{x} = \dfrac{21{,}000 + 26{,}000 + 31{,}000}{3} = \dfrac{78{,}000}{3} = \$26{,}000$; $s = 5{,}000$

    The mean will rise by \$1000 but, since the relative positions of the salaries is not changed, the standard deviation will not change.

    b. With 3% Raise: $\bar{x} = \dfrac{20{,}600 + 25{,}750 + 30{,}900}{3} = \dfrac{77{,}250}{3} = \$25{,}750$; $s = 5{,}150$

    Because 3% of each salary is a different amount, different amounts will be added to each salary so, both the mean and standard deviation will increase by 3%.

19. The mean will increase by the amount added but the range and standard deviation will not change. See problem 17a for an example.

21. Because everyone received the same raise, the standard deviation remains the same since the differences among the data as compared to the mean remain unchanged.

Practice Set 8-5

1. Since 830 hours is exactly one standard deviation above the mean (750 + 80 = 830 hours), based on the table of z-scores, 34.1% of the bulbs will last between 750 and 830 hours.

3. Since 670 hours is exactly one standard deviation below the mean (750 - 80 = 670 hours), based on the table of z-scores, 0.5 – 0.341 = 0.159 or 15.9% of the bulbs will burn out before being used 670 hours.

5. $z = \dfrac{15-15.5}{0.5} = \dfrac{-0.5}{0.5} = -1.00$; Based on the table of z-scores, 34.1% of values are between the mean (15.5 in) and 1s below average. To get the % of men that have a neck size of 15 in or larger, we add 34.1% to 50% (the percentage above the mean) to get a total of 84.1% of men with neck sizes greater than 15.5 in.

7. Using the percent calculated in problem 6, we know that 95.4% of men have neck sizes between 14.5 in. and 16.5 in. In a group of 75,000 men, we would expect that (95.4%)(75,000) = 0.954 × 75,000 = 71,550 of the men would have necks in this size range.

9. $z = \dfrac{350-500}{100} = \dfrac{-150}{100} = -1.50$; Based on the table of z-scores, 0.433 or 43.3% of values are between the mean and a z-score of -1.50. Therefore, 0.5 – 0.433 = 0.067 or 6.7% of values are below the z-score -1.50. Therefore, 6.7% of students are expected to score less than 350 on the SAT Math.

11. The area from the mean to the 0.01 area mark = 0.5 – 0.01 = 0.49. Look in the Table of z-Scores and locate the z-score associated with A = 0.490. You will find z = 2.31. Because we are looking at the lowest 1%, that is below the mean so the z-score will be negative. Substitute the known values into the z-score formula.
$$-2.31 = \dfrac{x-1200}{120}$$
$$(-2.31)(120) = x - 1200$$
$$-277.2 = x - 1200$$
$$922.8 = x$$
So 1% of the bulbs will burn out before approximately 920 hours of use.

13. Because the Table of z-scores gives areas from the mean to the z-score, if we want to find the area below 95%, we must find the z-score for the area that is between the mean and the upper 45% of the curve. Look in the Table of z-Scores and locate the z-score associated with A = 0.4500. You will find z = 1.64. Because this value is above the mean, the z-score will be positive. Substitute the known values into the z-score formula.
$$1.64 = \dfrac{x-1200}{120}$$
$$(1.64)(120) = x - 1200$$
$$196.8 = x - 1200$$
$$1396.8 = x$$
So 95% of the bulbs will burn out before approximately 1397 hours of use.

15. $z = \dfrac{22-20}{2.0} = \dfrac{2}{2} = 1.00$; Based on the table of z-scores, 0.341 or 34.1% of values are between the mean and this z-score. Therefore, 0.5 − 0.341 = 0.159 or 15.9% of values are above this score. Therefore, there is a 15.9% probability that a child will watch TV for more than 22 hours per week.

17. $z = \dfrac{15-20}{2.0} = \dfrac{-5}{2} = -2.50$ and $z = \dfrac{25-20}{2.0} = \dfrac{5}{2} = 2.50$. Based on the Table of z-Scores, the area between the mean and z = 2.50 is 0.494. Since the curve is symmetrical, the area between the mean and z = -2.50 is also 0.494, for a total proportion of 0.988 or about 99%. Therefore, there is a 99% probability that a child watches TV for between 15 and 25 hours per week.

19. $\bar{x} = \dfrac{37+52+25+48+26+41+22+15+52+40+50+16+58+59+51+26+39+12+37+20}{20} = \dfrac{726}{20} = 36.3$; $s = 15.10$

21. $z = \dfrac{65-72}{5} = \dfrac{-7}{5} = -1.40$; Based on the table of z-scores, 0.419 or 41.9% of values are between the mean and a z-score of -1.40. Therefore, 0.5 + 0.419 = 0.919 or 91.9% of values are above the z-score -1.40. Therefore, 91.9% of males are taller than 65 inches.

23. $z = \dfrac{64-72}{5} = \dfrac{-8}{5} = -1.60$ and $z = \dfrac{74-72}{5} = \dfrac{2}{5} = 0.40$. Based on the Table of z-Scores, the area between the mean and z = -1.60 is 0.445. The area between the mean and 0.40 is 0.155. Therefore, the total proportion between these two z-scores is 0.445 + 0.155 = 0.600 and 60% of males will have heights between 64 and 74 in.

25. Based on the definition of percentile, 85% will have scored lower than you.

27. Ray's $z = \dfrac{82-65}{12} = \dfrac{17}{12} = 1.42$; Susan's $z = \dfrac{78-65}{10} = \dfrac{13}{10} = 1.30$; Ray would receive a higher grade because his grade is farther above the mean average for his class than Susan's is for hers.

29. $z = \dfrac{82-78}{5} = \dfrac{4}{5} = 0.80$

31. A negative z-score indicates that the grade would be lower than the mean of 78.
$-1.6 = \dfrac{x-78}{5}$
$(-1.6)(5) = x - 78$

Fundamentals of Algebraic Modeling, 5th Ed

$$-8 = x - 78$$
$$70 = x \qquad \text{The grade would be 70.}$$

33. $z = \dfrac{110 - 100}{15} = \dfrac{10}{15} = 0.67$; The proportion of area between the mean and the z-score 0.67 is 0.249. Therefore, the percent of IQs below 110 will include the 50% below the mean of 100 and the 24.9% between 100 and 110 for a total of 74.9% or 75%.

35. The fifth percentile means that only 5% scored lower than this score. The z-score for 5% (0.050) below the mean = -1.64.
$$-1.64 = \dfrac{x - 100}{15}$$
$$(-1.64)(15) = x - 100$$
$$-24.6 = x - 100$$
$$75.4 = x \qquad \text{The IQ score of 75 corresponds to the 5}^{\text{th}} \text{ percentile.}$$

37. 95% in the middle means 47.5% above the mean and below the mean. Therefore, we look in the table for the z-score associated with an area (A) = 0.475. This value is 1.96. Since our two values will be both above and below the mean, we use +1.96 and -1.96 to find our answers.

$$1.96 = \dfrac{x - 100}{15} \qquad\qquad -1.96 = \dfrac{x - 100}{15}$$
$$(1.96)(15) = x - 100 \qquad\qquad (-1.96)(15) = x - 100$$
$$29.4 = x - 100 \qquad\qquad -29.4 = x - 100$$
$$129.4 = x \qquad\qquad 70.6 = x$$

The scores that separate the middle 95% from the remainder of the scores are 71 and 129.

39. $z = \dfrac{30 - 40}{11.4} = \dfrac{-10}{11.4} = -0.88$ and $z = \dfrac{50 - 40}{11.4} = \dfrac{10}{11.4} = 0.88$. Based on the Table of z-Scores, the area between the mean and z = -0.88 is 0.311. The area between the mean and 0.88 is also 0.311. Therefore, the total proportion between these two z-scores is 2(0.311) = 0.622. Therefore, 62.2% of customers use an ATM between 30 and 50 times. Among 5000 customers, we can predict that 62.2% of 5000 or 0.622 × 5000 = 3110 customers will use the ATM between 30 and 50 times.

41. $z = \dfrac{5\,ft.\,10\,in. - 5\,ft.\,5.5\,in.}{2.0} = \dfrac{4.5\,in.}{2.0} = 2.25$; The proportion of area between the mean and the z-score 2.25 is 0.488. Therefore, the percent of women's heights above $5\,ft.\,10\,in.$ will be calculated by taking the 50% above the mean of $5\,ft.\,5.5\,in.$ and subtracting the 48.8% from it for a difference of 1.2%.

Practice Set 8-6

1. The older the car, the less it is worth. This is a negative correlation.
3. As a car's weight increases, its gas mileage will decrease. This is a negative correlation.
5. a. Education is the independent variable and income is the dependent variable.
   b. There is a strong positive correlation based on the r-value.
   c. As one's years of education increase, one's income increases.
7. $y = 0.005(600) + 0.40 = 3.4$ GPA
9. $y = 0.002(1060) + 0.667 = 2.787$ GPA

11. a.

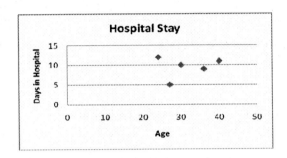

   b. There is no obvious linearity in the graph. The r-value (found using the calculator) is 0.158 which indicates weak correlation.
   c. $y = 0.065x + 7.35$ (found using the calculator)
   d. $y = 0.065(32) + 7.35 = 9.43$ or 9 days; This prediction is not reliable because of the weak correlation between these variables.

13. a.

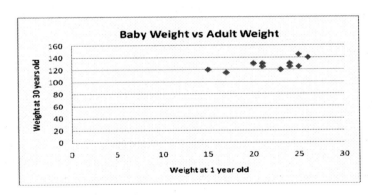

   b. There is a moderate correlation (r-value = 0.64).
   c. $y = 1.6x + 92.5$ (found using the calculator)
   d. The correlation coefficient is $r = 0.64$.
   e. $y = 1.6(20) + 92.5 = 124.5$ lb

15. a.

b. There seems to be a moderate correlation with a correlation coefficient $r = 0.55$. It appears in the graph to be a positive linear correlation.

17. a. Yes, there appears to be a positive linear correlation.

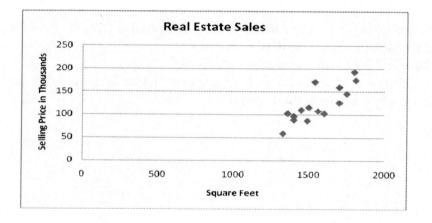

b. The correlation coefficient is $r = 0.82$ indicating a strong correlation.
c. $y = 0.197x - 185$ (Found using the calculator)
d. $y = 0.197(1575) - 185 = 125.275$ or \$125,275; This predicted price should be fairly reliable since the correlation between the two variables is strong.

Chapter 8 Review Problems

1. Descriptive statistics makes no predictions or guesses. It summarizes a population of data. Inferential statistics takes descriptive data and uses them to help make "educated guesses" about the data or about the group from which the data were gathered.

2. Quantitative data are data that can be "quantified," i.e., they have numbers and amounts associated with them. Qualitative data are non-numerical in nature.
3. cluster sample; probably not biased if flights chosen randomly from many airports
4. convenience sampling; could be biased if the interviewer chooses only people who seem friendly or who come in during a certain time of day
5. stratified sample; probably not biased if representative samples from each stratum are polled
6. Sample results cannot be applied to the general population since it was not representative.
7. The number seems too precise for a weight loss figure. Also, sample size is not given.
8. sum = 126, n = 9, mean = 14, median = 14, no mode, $R = 8$, $s = 2.7$ (all as determined using a graphing calculator and methods shown in the text)
9. sum = 173, n = 10, mean = 17.3, median = 8.5, mode = 8, $R = 55$, $s = 17.2$ (all as determined using a graphing calculator and methods shown in the text)
10. The data set in problem 9 has more variation since its standard deviation is larger.
11. For the data in problem 8, the mean is the best average because there are no big gaps in the list of numbers. For the data in problem 9, the median is the best average because of the relatively large jump in the data list to get to 57.
12.

Classes	Frequency
13-14	4
11-12	5
9-10	5
7-8	13
5-6	12
3-4	11

13. The histogram is positively skewed.

14. $z = \dfrac{37-35}{2} = \dfrac{2}{2} = 1.00$ The area between the mean and this z-score is 0.341. To find the area above this score, subtract 0.341 from 0.5 to give an area of 0.159 which indicates that about 16% of men have a waist size greater than 37 inches.

15. The mean average of the survey is 35 in. The mean average is the center of the normal distribution with 50% above and 50% below its value.

16. $z = \dfrac{37-35}{2} = \dfrac{2}{2} = 1.00$; $z = \dfrac{31-35}{2} = \dfrac{-4}{2} = -2.00$; The area between the mean and the z-score of 1.00 is 0.341. The area between the mean and the z-score of -2.00 is

0.477. To find the area between these two scores, add 0.341 and 0.477 to give an area of 0.818 which indicates that about 82% of men have a waist size between 31 inches and 37 inches.

17. math test: $z = \dfrac{88-80}{5} = \dfrac{8}{5} = 1.60$; history test: $z = \dfrac{88-85}{3} = \dfrac{3}{3} = 1.00$. The z-scores indicate that you scored farther above the class average on the math test than the history test. This means that, relatively speaking, you did better on the math test.

18. $z = \dfrac{30,000-45,000}{8200} = \dfrac{-15,000}{8200} = -1.83$; The area between the mean and this z-score is 0.466. To find the area below this score, subtract 0.466 from 0.5 to give an area of 0.034 which indicates the proportion that they expect to replace under this guarantee.

19. $z = \dfrac{40,000-45,000}{8200} = \dfrac{-5,000}{8200} = -0.61$; The area between the mean and this z-score is 0.229. To find the area below this score, subtract 0.229 from 0.5 to give an area of 0.271 which indicates the proportion that they expect to replace under this guarantee.

20. There does not appear to be a strong correlation.

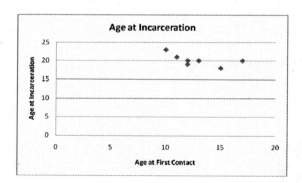

21. $y = -0.40x + 25.3$ (Found using the calculator)
22. $y = -0.40(14) + 25.3 = 19.7$ or about age 20
23. The correlation coefficient is $r = -0.61$ (Found using the calculator)
24. The r-value indicates a moderate negative correlation of these two variables.

Chapter 8 Test

1. quantitative because a *count* is a numerical amount
2. qualitative because an *occupation* is non-numerical
3. qualitative because the *name of a television* show is non-numerical
4. Because the number is a percent, you might investigate how many students took the test. It could be a very low number.
5. Data set B because of the unusually large value of 19 as compared to the other data

6. The mean is the best average for data set A, and the median is the best average for data set B because of the extreme value of 19 and how it affects the mean.

7. n = 11 and the data sum = 402, mean = $\frac{402}{11}$ = 36.5; median = 36 (the data are already sorted so count 6 items from either end); mode = 36; R = 44 - 32 = 12; s = 3.9 by the calculator method shown in the text

8. median = M

9.
```
0 | 3
1 | 2 7 0 8 8 9 2 8
2 | 7 7 2 9 1 6 1 8 9 1 6
3 | 0 5 8 6 5 8 2 3 2 1 2 3 1 2
4 | 2 7 1 5 3
5 | 1
```

10. It appears to be approximately normal.

11. $z = \frac{37-41}{4} = \frac{-4}{4} = -1.00$; The area between the mean and this z-score is 0.341. To find the area below this score, subtract 0.341 from 0.5 to give an area of 0.159 which indicates that about 16% of incoming students must take this course.

12. $z = \frac{65-80}{5} = \frac{-15}{5} = -3.00$

13. $1.75 = \frac{x-80}{5}$
    $(1.75)(5) = x - 80$
    $8.75 = x - 80$
    $88.75 = x$

14. $z = \frac{70-80}{5} = \frac{-10}{5} = -2.00$; The area between the mean and this z-score is 0.477. To calculate the entire area higher than the value of 70, add 0.477 + 0.500 to get 0.977 which is the proportion of students who passed.

15. $z = \frac{89-80}{5} = \frac{9}{5} = 1.80$; The area between the mean and this z-score is 0.464. This represents the proportion of students who scored between 80 (the mean) and 89.

16. A z-score is the number of standard deviations a value is above or below the mean average. The z-score for your height would be 2.00.

17. $z = \frac{60-64}{3} = \frac{-4}{3} = -1.33$; The area between the mean and this z-score is 0.408. To find the area below this score, subtract 0.408 from 0.5 to give an area of 0.092 which indicates the proportion that they expect to replace under this guarantee.

18. $y = 4(69)-130 = 146$ lb
19. $r = 0.80$ (Found using calculator); This value indicates a fairly strong correlation between an employee's age and his salary.

20.
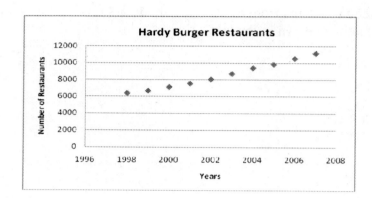

21. $y = 544.4x - 1081627.2$ (Found using the calculator)

22. $y = 544.4(2010) - 1,081,627.2 = 12616.8$ or 12617 predicted to be open in 2010